Fractional Discrete Chaos

Chaos

Theories, Methods and Applications

Topics in Systems Engineering

Print ISSN: 2810-9090
Online ISSN: 2810-9104

Series Editors: Luigi Fortuna *(Università degli Studi di Catania, Italy)*
Arturo Buscarino *(Università degli Studi di Catania, Italy)*

The Series aims to cover a wide spectrum of Engineering topics but with a strong characteristic of interdisciplinarity using less technical items. It could be a series including thinking aspects, history topics linked with engineering topics from civil engineering to industrial engineering and so on. The contributions will cover a wide series of engineering topics and could be useful both for educational purposes and for research information. The idea is to combine technology, science, arts and social science with some emerging subjects.

Published:

Topics in Systems Engineering — Volume 3

Fractional Discrete Chaos

Theories, Methods and Applications

Adel Ouannas

University of Larbi Ben M'hidi, Algeria

Iqbal M Batiha

Al Zaytoonah University of Jordan, Jordan

Viet-Thanh Pham

Ton Duc Thans University, Vietnam

NEW JERSEY · LONDON · SINGAPORE · BEIJING · SHANGHAI · HONG KONG · TAIPEI · CHENNAI · TOKYO

Published by

World Scientific Publishing Co. Pte. Ltd.

5 Toh Tuck Link, Singapore 596224

USA office: 27 Warren Street, Suite 401-402, Hackensack, NJ 07601

UK office: 57 Shelton Street, Covent Garden, London WC2H 9HE

Library of Congress Control Number: 2023004068

British Library Cataloguing-in-Publication Data
A catalogue record for this book is available from the British Library.

Topics in Systems Engineering — Vol. 3
FRACTIONAL DISCRETE CHAOS
Theories, Methods and Applications

ISBN 978-981-127-120-5 (hardcover)
ISBN 978-981-127-121-2 (ebook for institutions)
ISBN 978-981-127-122-9 (ebook for individuals)

For any available supplementary material, please visit
https://www.worldscientific.com/worldscibooks/10.1142/13277#t=suppl

Contents

Preface

Chaos is a phenomenon that arises in the natural world, from as tiny as a particle to as large as the universe. The theories of this phenomenon have essentially been included into all branches of science and engineering.

In electronic and electrical engineering, recent investigations have focused on a broad range of chaos theories including their stabilization, stimulation, analyses and their various applications. In fact, there are several books that handle many topics on chaos, concentrating on the mathematical formulation of systems' chaotic behaviors and their theoretical analyses. The primary aim of this book is to provide a comprehensive discussion on chaos by presenting the most recent researches related to fractional order discrete chaotic maps, illustrating advances and achievements related to these maps as well as some of their interdisciplinary applications, also presenting several theoretical tools associated with chaos, control and synchronization of discrete-time systems.

This book is one of the first that aims to cover the recent findings connected to fractional order chaotic discrete maps and their applications. It will be a useful reference in the field of chaos that can be generated by discrete fractional calculus.

Acknowledgment

The authors would like to thank Dr. Aicha Amina Khennaoui for her expertise and assistance throughout all aspects of our study and for her help in writing the book.

Chapter 1

Discrete Fractional Calculus

1.1. Introduction

Fractional calculus is an emerging field that deals with the investigation of fractional order integral and derivative, and their applications. More recently, the theory of fractional calculus has been used as a tool to study and model a variety of applied problems. In particular, it has been used extensively in the study of capacitor theory, electrical circuits, and from control theory to neuronal modeling [3, 9]. It is well known that the fractional derivative is widely applied to describe the regarding properties memory and history dependence of various materials and the processes including certain physical phenomena.

On the other hand, discrete fractional calculus is a very new area of science. This area was introduced in 1950. Starting from the idea of discretizing the Cauchy integral formula, Gray and Zhang in 1988 obtained the definition of Nabla approach while in 1989, Miller and Ross initiated the forward fractional differences and the fractional sum. Since then, quite a number of contributions on the topic have appeared, and many authors have studied the stability of fractional difference systems.

In the next sections, the Riemann–Liouville, Caputo and Grunwald–Letnikov fractional order difference operators and their properties are discussed. We see that the Caputo and the earlier defined Riemann–Liouville fractional discrete analogue are related to each other. The commutativity properties of the fractional sum and the fractional difference operators are given and discussed. As for Caputo fractional difference operator, we illustrate a very general Taylor difference formula. This is then applied to transform the initial value problem into an equivalent Voltera difference equation. Moreover, the h-difference operators and their properties are presented. Finally, Z-transform and *Laplace* methods in discrete fractional calculus are reported.

Numerous properties, and some not so elementary, are presented. Such inequalities are used to state and solve some initial value problems with the Riemann–Liouville and Caputo-like difference operators.

1.2. Preliminaries

We state here some important aspects and functions in the difference calculus that will be useful for our purposes. Let $a \in \mathbb{R}$, in the following, we restrict our attention to the time scale $\mathbb{N}_a = \{a, a + 1, a + 2 \ldots\}$.

Definition 1. Let $g : \mathbb{N}_a \to \mathbb{R}$, the forward difference operator Δ is defined by

$$\Delta g(t) = g(t+1) - g(t). \tag{1.1}$$

To compute the higher order differences we must proceed sequentially by composing the difference operator with itself. This means that, to compute the second order difference operator of the function $g(t)$, we have to use Eq. (1.1) twice. Hence, the second order difference is

$$\Delta^2 g(t) = \Delta g(t+1) - \Delta g(t) = g(t+2) - 2g(t+1) + g(t). \tag{1.2}$$

Proceeding similarly with (1.1), we obtain the nth order forward difference operator of the function $g(t)$ as follows

$$\Delta^n g(t) = \Delta \left(\Delta^{n-1} g(t) \right) = \sum_{k=0}^{n} \binom{n}{k} (-1)^{n-k} g(t+k), \tag{1.3}$$

where $t \in \mathbb{N}_a$ and $n = 1, 2, \ldots$. The notation $\binom{n}{k}$ is the binomial coefficient which is given as

$$\binom{n}{k} = \frac{n!}{k!(n-k)!}. \tag{1.4}$$

Now, we will present a few basic properties of the difference operator, which are very similar to the sum role, product rule, quotient role in differential calculus.

Theorem 1.1. [21]

- $\Delta\left(g(t) + g(t)\right) = \Delta g(t) + \Delta g(t).$
- *If K is a constant, $\Delta\left(Kg(t)\right) = K\Delta g(t).$*
- $\Delta\left(g(t)g(t)\right) = g(t)\Delta g(t) + g(t)\Delta g(t).$
- *For two positive integers m and n, $\Delta^m\left(\Delta^n g(t)\right) = \Delta^{m+n} g(t).$*
- $\Delta\left(\frac{g(t)}{g(t)}\right) = \frac{g(t)\Delta g(t) - g(t)\Delta g(t)}{g(t)g(t+1)}.$

Our next definition introduces Euler's Gamma function, which is considered as the most useful function in fractional calculus.

Definition 2. *Euler's Gamma function* is given by the integral

$$\Gamma(z) = \int_0^\infty t^{z-1} \exp^{-t} dt, \tag{1.5}$$

which is $\Gamma(\mu) > 0$ for $\mu > 0$.
One of the basic properties of the gamma function is

$$\Gamma(z) = z\Gamma(z),$$

with the normalizing condition $\Gamma(1) = 1$. Obviously, for $n \in \mathbb{N}_0$, $\Gamma(n+1) = n!$.

We shall also use the following function.

Definition 3. For arbitrary $\mu \in \mathbb{R}$, the falling factorial power function $s^{(n)}$, also denoted as $s^{\underline{\mu}}$, is defined as

$$s^{(\mu)} = \frac{\Gamma(s+1)}{\Gamma(s+1-\mu)}. \tag{1.6}$$

- In the case of positive integer value of μ, we have

$$s^{(\mu)} = s(s-1)(s-2)..(s-\mu+1).$$

where the product is zero if $s+1-j = 0$ for some j.
- In the case of negative integer value of μ, then we have

$$s^{(\mu)} = \frac{1}{(s+1)(s+2)\ldots(s-k)}.$$

According to the difference calculus, the following properties of the falling factorial power function are obtained as in [2].

Theorem 1.2. *Let μ be a real number, then we have*

- $\Delta s^{\underline{\mu}} = \mu s^{\underline{\mu-1}}$.
- *The generalized power rule*

$$\Delta(s-\alpha)^{\underline{\mu}} = \mu(s-\alpha)^{\underline{\mu-1}}. \tag{1.7}$$

- $(s-\mu)^{\underline{\mu}} = s^{\underline{\mu+1}}$, *where $\mu \in \mathbb{R}$.*
- $\mu^{\underline{\mu}} = \Gamma(\mu+1)$.
- *If $s \le r$, then $\forall \mu > r$; $s^{\underline{\mu}} \le r^{\underline{\mu}}$.*
- *If $0 < \mu < 1$ then $s^{\underline{a\mu}} \ge (s^{\underline{a\mu}})^{\mu}$, where $a \in \mathbb{R}$.*
- $s^{\underline{\alpha+\beta}} = (s-\beta)^{\underline{\alpha}} s^{\underline{\beta}}$.

1.3. Fractional Difference Operators

Fractional difference operator can be considered as a discretized version of the continuous fractional derivative. Different types of fractional difference operators have been proposed for different goals and application. Generally, there are four different definitions of fractional difference operators, they are, Grunwald–Letnikov, Caputo, Riemann–Liouville operators, and the fractional h-difference operators, which represent further generalizations of the fractional difference operators [3, 9]. To define these operators, we will first consider the definition of the fractional difference sum operator.

1.3.1. Fractional Difference Sum Operator

Let us consider the nth order initial value problem of the form

$$\begin{cases} \Delta^n x(t) = g(t), \\ x(a+i-1) = 0, \quad i = \overline{1,n}, \end{cases} \tag{1.8}$$

for $n \in \mathbb{N}$, $n > 0$. The solution of this IVP (1.8) is a function $x(t)$ such that (1.8) is satisfied. Similar to the indefinite integral of differential calculus, this solution is obtained by considering n-fold discrete integral of g as follows

$$x(t) = \Delta^{-n}g(t) = \sum_{s=a}^{t-1} \cdots \sum_{\tau_{n-1}=a}^{\tau_{n-2}-1} g(\tau_{n-1}) = \sum_{s=a}^{t-1} \frac{(t-\sigma(s))^{(n-1)}}{(n-1)!} g(s).$$

(1.9)

Thus, the n-fold sum of g is reduced into a single sum. Using the fact that $\frac{(t-s-1)^{(n-1)}}{(n-1)!} = 0$ for $s = t-1, t-2, \ldots, t-n+1$, we get to rewrite $\Delta^{-n}g(t)$ for $t \in \mathbb{N}_a$ as

$$\Delta_a^{-n}g(t) = \sum_{s=a}^{t-n} \frac{(t-\sigma(s))^{(n-1)}}{(n-1)!} g(s).$$

(1.10)

Clearly, the right-hand side of Eq. (1.10) is meaningful for any real number $\mu > 0$. The definition of the fractional sum of g of order μ is given next.

Definition 4. [4] For a function $g : \mathbb{N}_a \to \mathbb{R}$, the fractional difference sum of order $n - 1 < \mu \le n$ is given by

$$\Delta_a^{-\mu}g(t) = \frac{1}{\Gamma(\mu)} \sum_{s=a}^{t-\mu} (t-\sigma(s))^{(\mu-1)} g(s).$$

(1.11)

Remark 1. Since $\Delta_a^{-\mu}g$ is equal to zero under the assumption

$$\Delta_a^{-\mu}g(a+\mu-n) = \Delta_a^{-\mu}g(a+\mu-n+1) = \cdots = \Delta_a^{-\mu}g(a+\mu-1),$$

(1.12)

it is observed that $\Delta_a^{-\mu} : \mathcal{F}_{\mathbb{N}_a} \to \mathcal{F}_{\mathbb{N}_{a+\mu}}$, where $\mathcal{F}_{\mathbb{N}_a}$ and $\mathcal{F}_{\mathbb{N}_{a+\mu}}$ are the set of real-valued functions given of time scales $\mathbb{N}_a$ and $\mathbb{N}_{a+\mu}$, respectively.

The fractional difference sum can be given in terms of the binomial coefficient as follows

$$\Delta_a^{-\mu}g(t) = \sum_{k=0}^{-\mu+t-a} (-1)^k \binom{-\mu}{k} g(t-\mu-k).$$

(1.13)

Example 1. The fractional sum of the power rule function $g(t) = (t - a)^{(\alpha)}$ is equal to

$$\Delta_{a+\mu}^{-\mu}(t - a)^{(\alpha)} = \frac{\Gamma(\alpha + 1)}{\Gamma(\alpha + \mu + 1)}(t - a)^{(\alpha+\mu)}, \quad \forall t \in \mathbb{N}_{a+\alpha+\mu}, \quad (1.14)$$

where the proof is presented in [2, 6]. This formula implies that if $g(t) = K$, where K is a constant, then

$$\Delta_a^{-\mu} K = \frac{K}{\Gamma(\mu + 1)}(t - a)^{(\mu)}, \quad \forall t \in \mathbb{N}_{a+\mu}. \quad (1.15)$$

An important property of the fractional difference sum is given by the following theorem.

Theorem 1.3. [2] *Let g be a real-valued function defined on $\mathbb{N}_a$. If $\alpha > 0$ and $\mu > 0$, then the equation*

$$\Delta_{a+\alpha}^{-\mu}\Delta_a^{-\alpha}g(t) = \Delta_a^{-(\mu+\alpha)}g(t) = \Delta_{a+\mu}^{-\alpha}\Delta_a^{-\mu}g(t), \quad (1.16)$$

is satisfied at every point $t \in \mathbb{N}_{a+\alpha+\mu}$.

1.3.1.1. *Composing fractional sums and differences*

We now show how the fractional difference sum and the integer-order difference operator interact. Essentially, these results and their proofs have been discussed by Atici and Eloe in [3].

Proposition 1. [3] *Let $g : \mathbb{N}_a \to \mathbb{R}$ and let the fractional order $\mu > 0$. Then, the following equality holds*

$$\Delta^{-\mu}\Delta g(t) = \Delta\Delta^{-\mu}g(t) - \frac{(t - a)^{\mu-1}}{\Gamma(\mu)}g(a). \quad (1.17)$$

For $\mu = \mu + 1$ and using the fractional difference summation Eq. (1.17) leads us to the following Corollary.

Corollary 1. [3] *Assume g is defined on $\mathbb{N}_a$ and let $\mu > 0$, then*

$$\Delta_a^{-\mu-1}\Delta g(t) = \Delta_a^{-\mu}g(t) - \frac{(t - a)^{(\mu)}}{\Gamma(\mu + 1)}g(a). \quad (1.18)$$

This implies

$$\Delta^{-\mu}g(t) = \Delta^{-\mu-1}\Delta g(t) + \frac{(t-a)^{(\mu)}}{\Gamma(\mu+1)}g(a). \tag{1.19}$$

Let us now consider the case of an arbitrary positive integer-order.

Proposition 2. [3] *For any real number $\mu > 0$ and any positive integer q, the following equality holds*

$$\Delta^{-\mu}\Delta^q g(t) = \Delta^q\Delta^{-\mu}g(t) - \sum_{k=0}^{q-1}\frac{(t-a)^{\mu-q+k}}{\Gamma(\mu+k-q+1)}\Delta^k g(a), \tag{1.20}$$

where g is defined on $\mathbb{N}_a$.

If we replace $\mu = \mu + q$, Proposition 2 and Eq. (1.20) yield

$$\Delta^{-\mu}g(t) = \Delta^{-\mu-q}\Delta^q g(t) + \sum_{k=0}^{q-1}\frac{(t-a)^{(\mu+k)}}{\Gamma(\mu+k+1)}\Delta^k g(a). \tag{1.21}$$

Let us now consider the composition property of the fractional difference sum with the integer order difference operator.

Proposition 3. [3] *Let q be a positive integer and let $\mu > 0$ with $n-1 < \mu \le n$. Then for the function $f : \mathbb{N}_a \to \mathbb{R}$, we have*

$$\Delta^q[\Delta^{-\mu}g(t)] = \Delta^{-(\mu-q)}g(t), \tag{1.22}$$

for $t \in \mathbb{N}_{a+\mu}$.

1.3.1.2. *Leibenitz formula*

The Leibenitz formula is commonly used to compute the difference summation of products of two functions. The next theorem gives the fractional discrete version of the Leibenitz formula.

Theorem 1.4. [7] *Let $g(t)$ and $h(t)$ be defined on time scale $\mathbb{N}_0$ and $\mathbb{N}_a \cup \mathbb{N}_0$, respectively. Then for the fractional order $0 < \mu \le 1$*

$$\Delta_0^{-\mu}(gh)(t) = \sum_{q=0}^{\infty}\binom{-\mu}{q}[\Delta^q g(t)][\Delta_0^{-(\mu+q)}h(t+q)], \tag{1.23}$$

where $t \equiv \mu \, (mod \, 1)$ with $\binom{-\mu}{p}$ being the binomial coefficient which is equal to $\frac{\Gamma(-\mu+1)}{\Gamma(p+1)\Gamma(-\mu-p+1)}$.

1.3.2. Riemann–Liouville Difference Operator

Having established these fundamental definitions and properties of the fractional difference summation, we can extend the notion of the nth order difference operator to noninteger values of n. First, we give the definition of the Riemann–Liouville difference operator on time scale and present some of its properties.

Let g be a real-valued function defined on the time scale $\mathbb{N}_a$ and $n \in \mathbb{N}$. The Riemann–Liouville difference operator $^{RL}\Delta_a^\mu$ of fractional order $n - 1 < \mu < n$ is defined in [3]

$$
^{RL}\Delta_a^\mu g(t) = \Delta^n \Delta_a^{-(n-\mu)} g(t)
$$

$$
= \frac{1}{\Gamma(n-\mu)} \Delta^n \left(\sum_{s=a}^{t-(n-\mu)} (t - \sigma(s))^{(n-\mu-1)} g(t) \right). \quad (1.24)
$$

Remark 2. From the fractional difference sum domain it seems that the Riemann–Liouville difference operator $^{RL}\Delta_a^\mu g(t)$ is defined for $t = a + n \, mod(1)$.

The next theorem unifies the fractional difference operator.

Theorem 1.5. *Let $f : \mathbb{N}_a \to \mathbb{R}$ and $\mu > 0$ be given, with $n-1 < \mu \le n$. The following formula for the fractional difference is equivalent to (1.24)*

$$
^{RL}\Delta_a^\mu g(t) = \begin{cases} \dfrac{1}{\Gamma(-\mu)} \displaystyle\sum_{s=a}^{t+\mu} (t - \sigma(s))^{-\mu-1} g(s), & n - 1 < \mu < n, \\[6pt] \Delta^n g(t), & \mu = n. \end{cases}
$$

$$(1.25)$$

A proof of this theorem can be found in [6].

According to the definition of power rule function, the Riemann–Liouville difference operator given in Theorem 1.5 can be rewritten using the binomial coefficient as follows (see [6])

$$^{RL}\Delta_a^\mu g(t) = \sum_{p=0}^{\mu+t-a} (-1)^p \binom{\mu}{p} g(t+\mu-p), \quad t \in \mathbb{N}_{a+n-\mu}. \quad (1.26)$$

Remark 3. In the case of integer order, i.e. $\mu = n$, the Riemann–Liouville difference operator corresponds to Eq. (1.3).

Example 2. From the definition of the Riemann–Liouville difference operator, we derive

$$^{RL}\Delta^{\frac{4}{3}} g(t) = \Delta^2 \left(\Delta^{-\frac{2}{3}} t^{\frac{1}{3}} \right) = C\Delta^2 t^{\frac{2}{3}+\frac{1}{3}} = 0.$$

Example 3. As an example, we compute the Riemann–Liouville difference of the power rule function $g(t) = (t-a)^\alpha$ for $t \in \mathbb{N}_{a+\alpha+n-\mu}$

$$
\begin{aligned}
^{RL}\Delta_{a+\mu}^\mu (t-a)^{(\mu)} &= \Delta^n \left[\Delta_{a+\alpha}^{-(n-\mu)} (t-a)^{(\alpha)} \right] \\
&= \Delta^n \left[\frac{\Gamma(\alpha+1)}{\Gamma(\alpha+1+n-\mu)} (t-a)^{(\alpha+n-\mu)} \right] \\
&= \frac{\Gamma(\alpha+1)}{\Gamma(\alpha+1+n-\mu)} ((\alpha+n-\mu)\ldots(\alpha+1-\mu)) \\
&\quad \times (t-a)^{(\alpha-\mu)} \quad \text{using Eq. (1.14)} \\
&= \frac{\Gamma(\alpha+1)}{\Gamma(\alpha+1+n-\mu)} \frac{\Gamma(\alpha+n-\mu+1)}{\Gamma(\alpha+1-\mu)} (t-a)^{(\alpha-\mu)}.
\end{aligned}
$$

It is verified that the Riemann–Liouville operator of the power function $(t-a)^\alpha$ yields power functions of the same form.

In the last theorem of this subsection, we discuss another important property of fractional Riemann–Liouville difference operator, specifically the continuity of the operator with respect to the fractional order.

Theorem 1.6. [6] *Let* $g : \mathbb{N}_a \to \mathbb{R}$, $\mu > 0$ *and let* $t_{\mu,m} = a + \lceil \mu \rceil - \mu + m$ *be a fixed but arbitrary point in domain of operator* $\Delta_a^\mu f$ *where* $m \in \mathbb{N}_0$. *Then for each fixed* $m \in \mathbb{N}_0$, $\mu \to \Delta_a^\mu g(t_{t,m})$ *is continuous on* $[0, \infty)$.

1.3.2.1. *Composing with fractional sum and difference operators*

We indicate some properties of the Riemann–Liouville difference operator. First, we derive the following composition relations between fractional differences with fractional sums as proven in [6].

Proposition 4. [6] *Let us consider the fractional order* $\mu, \alpha > 0$ *with* $n - 1 < \mu \le n$. *Then*

$$^{RL}\Delta^{\mu}_{a+\alpha}\Delta^{-\alpha}_{a}g(t) = \Delta^{\mu-\alpha}_{a}g(t), \quad [t \in \mathbb{N}_{a+\alpha+n-\mu}]. \tag{1.27}$$

The following assertion shows that the left Riemann difference operator $^{RL}\Delta^{\mu}_{a+\mu}$ is an operation inverse to the μth fractional sum operator $\Delta^{-\mu}$ from the left and right.

Proposition 5. [8] *Let us now take* $\mu > 0$ *and* $g : \mathbb{N}_a \to \mathbb{R}$. *Taking into account that* $t \in \mathbb{N}_{a+n} \subset \mathbb{N}_a$, *we obtain*

$$^{RL}\Delta^{\mu}_{a+\mu}\Delta^{-\mu}g(t) = g(t), \tag{1.28}$$

and

$$\Delta^{-\mu}_{a+n-\mu}{}^{RL}\Delta^{\mu}g(t) = g(t), \quad [\mu \notin \mathbb{N}]. \tag{1.29}$$

Remark 4. Inequality (1.29) is not verified if the fractional order μ is not as required in the assumptions of Proposition 5.

In particular, if $t \in \mathbb{N}_{a+n-\alpha+\mu}$ we obtain

$$\Delta^{-\mu}_{a+n-\alpha}\Delta^{\alpha}_{a}g(t) = \Delta^{\alpha-\mu}_{a}g(t) - \sum_{k=0}^{n-1}\frac{\Delta^{k-(n-\alpha)}g(a+n-\alpha)}{\Gamma(\mu-n+k+1)}$$

$$\times (t-a-n+\alpha)^{(\mu-n+j)}, \quad [\alpha, \mu > 0, \, 0 < \mu \le n]. \tag{1.30}$$

Let us evaluate the difference of Riemann–Liouville operators of fractional order μ.

Proposition 6. [2] *For* $p - 1 < \mu < p$, *Proposition 1 implies that*

$$\Delta{}^{RL}\Delta^{\mu}g(t) = {}^{RL}\Delta^{\mu}\Delta g(t) + \frac{(t-a)^{(-\mu-1)}}{\Gamma(-\mu)}g(a). \tag{1.31}$$

Let us now consider the composition rule of two fractional difference operators ${}^{RL}\Delta^\mu ({}^{RL}\Delta^\alpha g(t))$. Unlike the integer order case, the Riemann–Liouville difference operator does not commute always, i.e. ${}^{RL}\Delta^\mu (\Delta^\alpha g(t)) \neq^{RL} \Delta^\alpha (\Delta^\mu g(t))$. To demonstrate this, we start with the commutativity property of the Riemann–Liouville difference operator with the integer order difference operator.

Lemma 1. *For any integer number q and for the fractional order $\mu > 0$ with $n - 1 < \mu \leq n$, we have*

$$\Delta^q \, {}^{RL}\Delta_a^\mu g(t) = \Delta_a^{q+\mu} g(t), \quad [\forall n \in \mathbb{N}, \forall t \in \mathbb{N}_{a+n-\mu}]. \tag{1.32}$$

The following index rule property is also easy to prove.

Theorem 1.7. [6] *Let $\mu, \alpha > 0$ and $n, m \in \mathbb{N}$ where $n - 1 < \mu \leq n$ and $m - 1 < \alpha \leq m$. The following equality*

$$\Delta_{a+m-\mu}^\mu \Delta_a^\mu g(t) = \Delta^{\mu+\mu} g(t) - \sum_{j=0}^{m-1} \frac{\Delta_a^{j-(n-\mu)} g(a + m - \mu)}{\Gamma(-\mu - m + j + 1)}$$

$$\times (t - a - m + \mu)^{-\mu-m+j}, \tag{1.33}$$

holds for $t \in \mathbb{N}_{a+m \ \mu+n-\alpha}$. By the same token as in Theorem 1.7, we may write in reverse order

$${}^{RL}\Delta_{a+n-\alpha}^\mu \, {}^{RL}\Delta_a^\alpha g(t) = {}^{RL}\Delta^{\mu+\alpha} g(t) - \sum_{j=0}^{n-1} \frac{\Delta_a^{j-(m-\alpha)} g(a + m - \mu)}{\Gamma(-\mu - n + j + 1)}$$

$$\times (t - a - n + \mu)^{-\mu-m+j}, \tag{1.34}$$

where the terms in the summation vanish if $\mu \in \mathbb{N}_0$.

Now we present the rules for fractional integration by parts, which were proved in [7].

Theorem 1.8. [7] *Let $g(t)$ and $h(t)$ be defined on the time scale $\mathbb{N}_a$. Then for the fractional order $0 < \mu < 1$*

$$\sum_{t=a}^{b-1} g(t + \mu - 1)\,{}^{RL}{}_t\Delta_{a+\mu-1}^{\mu} h(t)$$

$$= \sum_{t=a}^{b-1} h(t + \mu - 1)_{b+\mu-1}{}^{RL}\Delta_s^{\mu} g^{\rho}(t + 2(\mu - 1)), \quad (1.35)$$

where $g^{\rho} = g \circ \rho$ with $\rho(t) = t - 1$.

1.3.2.2. *Initial value problem*

In this subsection, we discuss an initial value problem for orders $\mu \in]0, 1]$ and obtain the existence and uniqueness of a solution. Atici and Eloe established the following result [3].

Let $g : \mathbb{N}_0 \times \mathbb{R} \to \mathbb{R}$ and $\mu \in [0, 1]$. Consider the following nonlinear fractional difference equation with an initial condition

$$\begin{cases} {}^{RL}\Delta_a^{\mu} x(t) = g(t + \mu - 1, x(l + \mu - 1)), & t \in \mathbb{N}_0, \\ \Delta^{\mu-1}\big|_{t=0}\, x = x_0, \end{cases} \quad (1.36)$$

where x_0 is a real number. The solution, $x(t)$, if it exists, is defined on $\mathbb{N}_{\mu-1}$.

We apply the fractional difference summation $\Delta^{-\mu}$ to Eq. (1.36) and from Theorem 1.3 we get

$$x(t) = \frac{t^{(\mu-1)}}{\Gamma(\mu)} a_0 + \frac{1}{\Gamma(\mu)} \sum_{s=0}^{t-\mu} (t - \sigma(s))^{(\mu-1)}$$

$$\times\, g(s + \mu - 1, x(s + \mu - 1)), \quad t \in \mathbb{N}_{\mu-1}. \quad (1.37)$$

The recursive iteration to this sum equation implies that (1.37) is representing the unique solution of the initial value problem (1.36).

Example 4. [3] Define

$$\begin{cases} {}^{RL}\Delta_a^{\mu} x(t) = \lambda x(t + \mu - 1), & [t \in \mathbb{N}_0], \\ \Delta^{\mu-1}\big|_{t=0} = x_0 = x(\mu - 1). \end{cases} \quad (1.38)$$

After applying the fractional difference summation $\Delta^{-\mu}$ to Eq. (1.38) as described in (1.37), we have

$$y(t) = \frac{t^{(\mu-1)}}{\Gamma(\mu)} a_0 + \frac{\mu}{\Gamma(\mu)} \sum_{s=0}^{t-\mu} (t - \sigma(s))^{(\mu-1)} y(s + \mu - 1). \qquad (1.39)$$

After applying the method of successive approximations, set

$$x_0(t) = \frac{t^{(\mu-1)}}{\Gamma(\mu)} x_0,$$

$$x(t) = x_0(t) + \frac{\mu}{\Gamma(\mu)} \sum_{s=0}^{t-\mu} (t - \sigma(s))^{(\mu-1)} x_{m-1}(s+\mu-1) = x_0(t) + \lambda \Delta^{-\mu} x_{m-1}(t+\mu-1).$$

Using the the power rule function we get

$$x(t) = x_0(t) + \lambda \Delta^{-\mu} x_0(t + \mu - 1) = x_0(t) + \left(\frac{t^{(\mu-1)}}{\Gamma(\mu)} + \lambda \frac{(t+\mu-1)^{(2\mu-1)}}{\Gamma(2\mu)} \right).$$

With repeated applications of the power rule, it follows inductively that

$$x_m(t) = x_0(t) + \sum_{i=0}^{m} \frac{\lambda^i}{\Gamma((i+1)\mu)} (t + (i-1)(\mu-1))^{(i\mu+\mu-1)}, \quad m = 0, 1, 2, \ldots$$

Formally, take the limit $m \to \infty$ to obtain

$$x_m(t) = x_0(t) + \sum_{i=0}^{\infty} \frac{\lambda^i}{\Gamma((i+1)\mu)} (t + (i-1)(\mu-1))^{(i\mu+\mu-1)},$$

$$m = 0, 1, 2, \ldots \qquad (1.40)$$

One immediate observation can be made. Set $\mu = 1$. Then $x(t) = x_0 \sum_{i=0}^{\infty} \frac{\lambda^i}{i!} t^{(i)}$. Since the initial value problem with $\mu = 1$ has the unique solution $x_0(1 + \lambda)t$, we obtain $(1 + \lambda)^t = \sum_{i=0}^{\infty} \frac{\lambda^i}{i!} t^{(i)}$.

1.3.3. *Caputo Fractional Difference Operator*

In this section, we present the definitions and some properties of the left-Caputo fractional difference of order μ. The main advantage of Caputo's approach is that the initial condition for fractional difference equations with fractional left-Caputo difference operator takes on the same form as for integer order difference equations, unlike the Riemann–Liouville difference approach which leads to initial conditions containing limit values.

Let us assume that $n - 1 < \mu < n$, where n denotes a positive integer, and that $n = [\mu] + 1$. Caputo-like difference operator definition can be written as [9]

$$
{}^{C}\Delta_a^\mu = \Delta_a^{-(n-\mu)}\Delta^n g(t)
$$

$$
= \frac{1}{\Gamma(n-\mu)} \sum_{s=a}^{t-(n-\mu)} (t - \sigma(s))^{(n-\mu-1)} (\Delta_s^n f)(s), \quad \forall t \in \mathbb{N}_{a+\mu}.
$$

$$(1.41)$$

Obviously, if μ is a positive integer ($\mu = n$), then Eq. (1.41) reduces to the integer-order difference formula

$$
\Delta^n g(t) = \sum_{r=0}^{n} (-1)^{r+1} \binom{n}{r} g(r+k).
$$

$$(1.42)$$

From the relation (1.42), the author in [10] established an equivalent formula for the fractional left-Caputo difference of order μ, its use being essential in many applications.

Proposition 7. *Let $n - 1 < \mu \le n$ and set $\mu = n - \mu$, where $n \in \mathbb{N}$. The following formula is equivalent to (1.41):*

$$
{}^{C}\Delta_a^\mu g(t) =
\begin{cases}
\dfrac{1}{\Gamma(n-\mu)} \displaystyle\sum_{s=a}^{t-(n-\mu)} (t - \sigma(s))^{(n-\mu-1)} \\
\qquad \times \displaystyle\sum_{r=0}^{n} (-1)^{r+1} \binom{n}{r} g(r+k), & \mu \in (n-1, n) \\
\Delta^n g(t), & \mu = n.
\end{cases}
$$

$$(1.43)$$

Example 5. [9] We begin with the simple function $g(t) = K$, where K is a constant. Thus, from (1.41)

$$
{}^{C}\Delta_a^\mu K = 0, \quad \mu > 0, \quad a \in \mathbb{R}.
$$

$$(1.44)$$

This formula is straight-forward and may be proved by direct evaluation.

Example 6. [9] Let us now apply the left-Caputo difference operator to $g(t) = (t-a)^{(\alpha)}$. One could have

$$\Delta_{a+\mu}^{\mu}(t-a)^{(\alpha)} = \frac{\Gamma(\alpha+1)}{\Gamma(\alpha-\mu+1)}(t-a)^{(\alpha-\mu)}, \quad t \in \mathbb{N}_{a+\alpha+n-\mu}, \quad (1.45)$$

where $n = \lceil\mu\rceil$ and $\mu > n$. One may obtain (1.45) by direct generalization of Δ^n difference of f. The classical method of proving that is as follows. First note that $\Delta(t-a)^{(\mu)} = \mu(t-a)^{(\mu-1)}$. Furthermore,

$$\Delta^n(t-a)^{(\alpha)} = \alpha(\alpha-1)\ldots(\alpha-(n-1))(t-a)^{(\alpha-n)},$$
$$= \frac{\Gamma(\alpha+1)}{\Gamma(\alpha-n+1)}(t-a)^{(\alpha-n)}.$$

Introducing the Caputo-like delta difference defined in (1.41) leads to

$$\begin{aligned}
\Delta^{\mu}(t-a)^{(\alpha)} &= \Delta_{a+\mu}^{-(n-\mu)}\Delta^n(t-a)^{(\alpha)} \\
&= \frac{\Gamma(\alpha+1)}{\Gamma(\alpha-n+1)}\Delta^{-(n-\mu)}(t-a)^{(\mu-n)} \\
&= \frac{\Gamma(\alpha+1)}{\Gamma(\alpha-n+1)}\frac{\Gamma(\alpha-n+1)}{\Gamma(\mu-\mu+1)}(t-a)^{(\alpha-\mu)},
\end{aligned}$$

where $t \in \mathbb{N}_{a+\mu-n+\mu}$.

Let g be a real-valued function defined on the time scale $\mathbb{N}_a$, and let $\left(^{RL}\Delta_a^{\mu}g\right)(t)$ be the Riemann–Liouville difference operator. The relationship between the Riemann–Liouville and the left-Caputo difference operators for any $\mu > 0$ is given as [9]

$$^C\Delta_a^{\mu}g(t) = {}^{RL}\Delta_a^{\mu}g(t) - \sum_{k=0}^{n-1}\frac{(t-a)^{(k-\mu)}}{\Gamma(k-\mu+1)}\Delta^k g(a). \quad (1.46)$$

1.3.3.1. *Properties of fractional left-Caputo difference operator*

Analogue to the properties of the Riemann–Liouville difference operator, the left-Caputo fractional difference operator $\left(^C\Delta_a^{\mu}g\right)(t)$ provides operations inverse to the μ-fractional difference summation from the right when $0 < \mu \leq 1$. But they do not have such properties in the general case. The following proposition is valid.

Proposition 8. [10] *Let $f : \mathbb{N}_a \to \mathbb{R}$.*

- *If $0 < \mu \leq 1$, then*

$$\Delta_{a+(n-\mu)}^{-\mu}\,{}^C\Delta_a^\mu g(t) = g(t) - g(a). \tag{1.47}$$

- *If $\mu > 0$, then*

$$\Delta_{a+(n-\mu)}^{-\mu}\,{}^C\Delta_a^\mu g(t) = g(t) - \sum_{k=0}^{n-1} \frac{(t-a)^k}{k!}\Delta^k g(a). \tag{1.48}$$

The next theorem yields the summation by parts formula of the left-Caputo difference operator.

Theorem 1.9. [10] *Let us take functions $g(t)$ and $h(t)$, defined on $\mathbb{N}_a \cap b_{\mathbb{N}}$ where $a \equiv b \ (mod\ 1)$. Then*

$$\sum_{s=a+1}^{b+1} h(s)\,{}^C\Delta_a^\mu g(s-\mu) = g(s)_{b-1}\Delta^{-(1-\mu)}h(s-(1-\mu))|_a^{b-1}$$

$$+ \sum_a^{b-2} g(s)\,{}^C_{b-1}\Delta^\mu h(s+\mu). \tag{1.49}$$

1.3.3.2. *Taylor difference formula*

Another important basic result in classical analysis is Taylor's theorem. In the following, we will introduce a new generalization of Taylor's formula that involves Caputo-like difference operator along with its proprieties [11].

The classical Taylor difference formula of elementary calculus is [12]

$$g(t) = \sum_{i=0}^{n-1} \frac{(t-a)^{(i)}}{i!}\Delta^i g(a) + \frac{1}{(n-1)!}\sum_{l=a}^{t-n}(t-l-1)^{(n-1)}\Delta^n g(l). \tag{1.50}$$

Now we extend the Taylor formula (1.50) to fractional Caputo-like difference operator. The importance of this formula lies in its help in the analysis of the initial value problem.

Theorem 1.10. [11] *Suppose that g is a real function defined on $\mathbb{N}_a$ with $a \in \mathbb{Z}^+$. Then, for $\mu > 0$ and $\mu = m - \mu$, the fractional Caputo-like difference operator*

$$g(t) = \sum_{k=0}^{m-1} \frac{(t-a)^k}{k!} \Delta^k g(a) + \frac{1}{\Gamma(\mu)} \sum_{s=a+\mu}^{t-\mu} (t-s-1)^{(\mu-1)} \Delta^\mu g(s),$$

(1.51)

exists for all $t \in \mathbb{N}_{a+m}$. In particular, if $0 < \mu < 1$, then the Caputo generalized Taylor's formula (1.51) *reduces to*

$$g(t) = g(0) + \frac{1}{\Gamma(\mu)} \sum_{s=1-\mu}^{t-\mu} (t-s-1)^{(\mu-1)} \Delta^\mu g(s).$$

(1.52)

Remark 5. Let g be a function on $[a, b] = [a, a+1, a+2, \ldots, b]$. Then clearly, for $n - 1 < \mu < n$, the fractional discrete Taylor's formula (1.51) is only valid for $t \in [a + n, b]$, $a + n < b$.

Using the fact that

$$\Delta^q \left(\frac{(t-a)^k}{k!} \right) = \frac{(t-a)^{k-q}}{(k-q)!},$$

for every integer order q, the Taylor's formula is extended to:

$$\Delta^q g(t) = \sum_{k=q}^{n-1} \frac{(t-a)^{(k-p)}}{(k-q)!} \Delta^k g(a)$$

$$+ \frac{1}{\Gamma(\mu - p)} \sum_{s=a+\mu}^{t-\mu+q} (t-s-1)^{(\mu-q-1)} \Delta_*^\mu g(s), \quad (1.53)$$

$\forall t \in \mathbb{N}_{a+n-q}$.

It can be noticed that if $q = 0$, $\Delta^k g(a) = 0$, for $k = \overline{0, n-1}$, Eq. (1.53) is reduced to

$$g(t) = \frac{1}{\Gamma(\mu)} \sum_{s-a+\mu}^{t-\mu} (t-s-1)^{(\mu-1)} \Delta_*^\mu g(s), \quad \forall t \in \mathbb{N}_{a+n}. \quad (1.54)$$

1.3.3.3. *Initial value problem*

This subsection is devoted to illustrate the theorem of existence and uniqueness of solutions to the left-Caputo difference equations with fractional order and initial conditions. Most of the investigations in this field involve the existence and uniqueness of solutions to fractional difference equations with the left-Caputo operator having the form

$$\begin{cases} {}^{C}\Delta_a^{\mu} y(t) = g(t-1+\mu, y(t-1+\mu)), & t \in \mathbb{N}_{a+1-\mu}, \\ y_0(t) = y(a), \end{cases} \quad (1.55)$$

where $0 < \mu \leq 1$, g is the continuous function defined by $g : [0, +\infty) \times \mathbb{R} \rightarrow \mathbb{R}$. Fulai *et al.* in [13] proved the following theorems.

Theorem 1.11. [13] *The function $y(t)$ is a solution of the initial value problem (1.55) if and only if it is a solution of the following discrete Volterra equation*

$$y(t) = y_0(t) + \frac{1}{\Gamma(\mu)}$$

$$\times \sum_{s=a+1-\mu}^{t-\mu} (t-s-1)^{(\mu-1)} g(s+\mu-1, y(s+\mu-1)), \quad (1.56)$$

where $0 < \mu \leq 1$ and $t \in \mathbb{N}_{a+1}$.

Theorem 1.12. [13] *Let $(X, ||.||)$ be real Banack space where $||y|| = \sup\{||x(t)||, t \in \mathbb{N}_a\}$, and assume that the function g is continuous in X and it is globally Lipshitz with constant L. Then, the IVP (1.55) has a unique solution $y(t)$ provided that $0 < L < \frac{1}{1+\mu}$.*

1.4. Other Fractional Difference Operators

1.4.1. *Grunwald–Letnikov Fractional Difference Operator*

In this subsection, we give the definition of the fractional order Grunwald–Letnikov difference operator.

Definition 5. The *fractional order Grunwald–Letnikov* is given by

$$\Delta^\mu g(k) = \frac{1}{h^\mu} \sum_{j=0}^{k} (-1)^j \binom{\mu}{j} g(k-j), \qquad (1.57)$$

where the fractional order $\mu \in \mathbb{R}^{*+}$, i.e. the set of strictly positive real numbers, $h \in \mathbb{R}^{*+}$ is a sampling time taken equal to unity in all what follows, and $k \in \mathbb{N}$ represents the discrete time. The term $\binom{\mu}{j}$ can be obtained from the following relation:

$$\binom{\mu}{j} = \begin{cases} 1 & \text{for} \quad j = 0, \\ \dfrac{\mu(\mu-1)\ldots(\mu-j+1)}{j!} & \text{for} \quad j > 0. \end{cases} \qquad (1.58)$$

1.4.2. *Fractional h-Difference Operators*

In this section, some basic concepts related to the fractional h-difference operators are briefly summarized. In this context, we first introduce some definitions and notation [14]. Let $h > 0$ and take $(h\mathbb{N})_a = \{a, a+h, a+2h, \ldots\}$ with $a \in \mathbb{R}$.

Definition 6. Let h be a strictly real positive number and $g : (h\mathbb{N})_a \to \mathbb{R}$, the forward h-difference operator is introduced as:

$$\Delta_h g(t) = \frac{g(\sigma_h(t)) - g(t)}{h}, \qquad (1.59)$$

where $\sigma(sh) = (t+1)h$.

Definition 7. The *h-falling factorial function* of real order μ is defined as

$$t_h^{(\mu)} = h^\mu \frac{\Gamma\left(\frac{t}{h}+1\right)}{\Gamma\left(\frac{t}{h}+1-\mu\right)}, \qquad (1.60)$$

in which $t \in \mathbb{R}$. As one can see when $h = 1$, the h-falling factorial function is equivalent to the falling factorial function defined by Eq. (1.6). One also expects to see that $t_h^{(\mu)}$ converges to $t^{(\mu)}$ when h tends to zero. This is illustrated in the next proposition.

Proposition 9. [15] *For $t \geq 0$ and $\mu \in \mathbb{R}$,*

$$\lim_{h \to 0} t_h^{(\mu)} = t^{(\mu)}. \tag{1.61}$$

Definition 8. [15] Let $g : (h\mathbb{N})_a \to \mathbb{R}$ and $0 < \mu$ be given. a is a starting point. The μth order *fractional h-summation* is given as

$$_h\Delta_a^{-\mu}g(t) = \frac{h}{\Gamma(\mu)} \sum_{s=\frac{a}{h}}^{\frac{t}{h}-\mu} (t - \sigma(sh))_h^{(\mu-1)} g(sh),$$

$$\sigma(sh) = (s+1)h, \ a \in \mathbb{R}, \ t \in (h\mathbb{N})_{a+\mu h}, \tag{1.62}$$

where the h-falling factorial function implies $(h\mathbb{N})_{a+(1-\mu)h} = \{a + (1 - \mu)h, a + (2 - \mu)h, \dots\}$.

Remark 6. The fractional h-difference summation $_h\Delta_a^{-\mu}$ maps every function defined from $(h\mathbb{N})_a$ into the function defined as $(h\mathbb{N})_{a+\mu h}$.

Accordingly to the definition of h-factorial function, the formula given in Definition 8 can be rewritten as

$$\left(_h\Delta_a^{-\mu}g\right)(t) = h^\mu \sum_{k=0}^{n} \frac{\Gamma(\mu + n - k)}{\Gamma(\mu)\Gamma(n - k + 1)} g(a + kh)$$

$$= h^\mu \sum_{k=0}^{n} \binom{n - k + \mu - 1}{n - k} g(a + kh) \tag{1.63}$$

$$= h^\mu \sum_{j=0}^{n} (-1)^j \binom{-\mu}{j} g(\mu - jh),$$

for $t = a + (\mu + n)h, \ n \in \mathbb{N}_0$.

Next, the h-power rule and the commutative property of the fractional h-difference summation operator are reported. The proof for the case $h > 0$ can be found in [16].

Lemma 2. *Let μ be a real number such that, can be found*

$$_a\Delta_h^{-\mu}(t - a + \mu h)_h^{(\mu)} = \frac{\Gamma(\mu + 1)}{\Gamma(\mu + \mu - 1)}(t - a + \mu h)_h^{(\mu+\mu)}. \tag{1.64}$$

Proposition 10. *Let f be a real-valued function defined on $(h\mathbb{N})_a$, where $a, h \in \mathbb{R}$, $h > 0$. For $\mu, \mu > 0$ the following equalities hold*

$$_h\Delta_{a+\mu h}^{-\mu} {}_h\Delta_a^{-\mu} g(t) = {}_h\Delta_a^{-(\mu+\mu)} g(t) = {}_h\Delta_{a+\mu h}^{-\mu} {}_h\Delta_a^{-\mu} g(t),$$

$$t \in (h\mathbb{N})_{a+(\mu+\mu)h}. \qquad (1.65)$$

Based on the above definition of the h-fractional sum, it is possible to define the fractional h-difference operators.

Definition 9. Let $\mu \in (0, 1]$, the *Riemann-like fractional h-difference operator*

$$_a^{RL}\Delta_h^\mu g(t) = \left(\Delta_h \left({}_h\Delta_a^{-(1-\mu)} f \right) \right)(t), \quad t \in (h\mathbb{N})_{a+(1-\mu)h}. \qquad (1.66)$$

Remark 7. Note that the the *Riemann-like fractional h-difference operator* maps every function defined from $(h\mathbb{N})_a$ into $(h\mathbb{N})_{a+(1-\mu)h}$.

Definition 10. [9] For $f(t)$ defined on $(h\mathbb{N})_a$ and $0 < \mu$, $\mu \notin \mathbb{N}$, the Caputo-like difference is defined as

$$_h^C\Delta_a^\mu g(t) = \Delta_a^{-(n-\mu)}\Delta^n g(t), \quad t \in (h\mathbb{N})_{a+(n-\mu)h}, \qquad (1.67)$$

where $n = \lceil \mu \rceil + 1$. If $\mu = 1$, then we have:

$$_h^C\Delta_a^\mu g = \Delta^m g, \quad t \in (h\mathbb{N})_a. \qquad (1.68)$$

The last operator that we take under our consideration is fractional h-difference Grunwald–Letnikov-like operator.

Definition 11. Let $\mu \in \mathbb{R}$, the *Grunwald–Letnikov-like h-difference operator* Δ_h^μ of order μ for a function $g : (h\mathbb{N})_a \to \mathbb{R}$ is defined as

$$\Delta_h^\mu g(t) = \frac{1}{h^\mu} \sum_{s=0}^{\frac{t-a}{h}} (-1)^s \binom{\mu}{j} g(t - sh), \qquad (1.69)$$

where the term $\binom{\mu}{s}$ can be obtained from the following relation

$$\binom{\mu}{s} = \begin{cases} 1 & \text{for} \quad s = 0, \\ \dfrac{\mu(\mu - 1)\ldots(\mu - s + 1)}{s!} & \text{for} \quad s \in \mathbb{N}. \end{cases} \qquad (1.70)$$

Remark 8. The *Grunwald–Letnikov-like h-difference operator* maps every function defined from $(h\mathbb{N})_a$ into the same domain $(h\mathbb{N})_a$.

1.5. Transform Methods

1.5.1. *Z-Transform Method*

The *Z*-transform is an alternative method which is used to solve linear difference equations as well as certain summation equations. Here, we attempt to review the *Z-transform* with fractional order.

Definition 12. [17] The *Z*-transform of a sequence $\{y_k\}$ of a complex variable is defined as

$$Z(x(n)) = \sum_{j=0}^{\infty} y(j) z^{-j}, \tag{1.71}$$

where $y(n) = 0$ for $n \in \mathbb{N}$.

With the help of using the ratio test $\lim_{j\to\infty} \left| \frac{y(j+1)}{y(j)} \right| = R$, it is found that the *Z*-transform converges if

$$\lim_{j\to\infty} \left| \frac{y(j+1) z^{-j-1}}{y(j) z^{-j}} \right| < 1.$$

Hence the series (1.71) converges in the region $|z| > R$ and diverges for $|z| < R$.

Let us now evaluate a *Z*-transform of the *h*-fractional difference summation $\Delta_h^{-\mu}$. To establish this, let us write $\Delta_h^{-\mu}$ in the form of the convolution of $\varphi_\mu(n) = \binom{n+\mu-1}{n}$ and $\overline{y}(s) = y(a+sh)$, for $t = a + (\mu+n)h$, as

$$\left({}_a\Delta_h^{-\mu} y \right)(t) = h^\mu \left(\varphi_\mu * y(a+sh) \right)(n) = \sum_{s=0}^{n} \binom{n-s+\mu-1}{n-s} \overline{y}(s). \tag{1.72}$$

Using the fact that $\varphi_\mu(n) = 0$ with $n < 0$, and for $n \in \mathbb{N}_0$

$$\varphi_\mu(n) = \binom{n+\mu-1}{n} = (-1)^n \binom{-\mu}{n},$$

the Z-transform of the fractional difference summation is written as the following proposition.

Proposition 11. [18] *Let* $t = a + \mu h + nh \in (h\mathbb{Z})_{a+\mu h}$, *then the Z transform of* $\left(_a\Delta_h^{-\mu}y \right)(t)$ *is given as*

$$\mathcal{Z}\left(_a\Delta_h^{-\mu}y \right)(z) = \left(\frac{hz}{z-1} \right)^\mu X(z), \tag{1.73}$$

where $X(z) = \mathcal{Z}[\bar{x}](z)$.

Next, we define the Z-transform of the Riemann–Liouville fractional difference operator. First, we define the family of functions $\varphi_{k,\mu} : \mathbb{Z} \to \mathbb{R}$ parametrized by $k \in \mathbb{N}_0$ and $\mu \in (0,1]$ with the following values

$$\varphi_{k,\mu} = \begin{cases} \dbinom{n-k+k\mu+\mu-1}{n-k} & \text{for } n \in \mathbb{N}_k, \\ 0 & \text{for } n < k. \end{cases} \tag{1.74}$$

The Z-transform of the Riemann–Liouville h-difference operator is composed as follows.

Proposition 12. *Let* $t \in (h\mathbb{N})_{a+(1-\mu)h}$, *and* $\mu \in (0,1]$, *then*

$$\mathcal{Z}\left(_a^{RL}\Delta_h^\mu x \right)(z) = z\left(\frac{hz}{z-1} \right)^{-\mu} \mathcal{Z}[y(a+nh)](z) - zh^{-\mu}y(a). \tag{1.75}$$

Let us introduce the following notation for the left-Caputo h-difference operator using the binomial function φ

$$\left(_a^C\Delta_h^\mu y \right)(t) = h^{-\mu}\left(\varphi_{1-\mu} * \Delta y(a+nh) \right)(n),$$
$$[t = a + (1-\mu)h + nh]. \tag{1.76}$$

Proposition 13. *For the fractional order* $\mu \in (0,1]$ *and* $t \in (h\mathbb{N})_{a+(1-\mu)h}$, *the Z-transform of the left-Caputo h-difference* *operator*

$$\mathcal{Z}\left({}^{C}_{a}\Delta^{\mu}_{h}y\right)(t)(z) = h^{-\mu}\left(\frac{z}{z-1}\right)^{1-\mu}\left((z-|1)X|(z) - zy(a)\right),$$

$$(1.77)$$

where $X(z) = \mathcal{Z}[y(a+nh)]$.

The next step is to present $\mathcal{Z}$-transform of the Grunwald–Letnikov h-difference operator.

Proposition 14. *For the fractional order* $\mu \in (0,1]$, *let us define* $y(n) = (t)$, *where* $t \in (h\mathbb{N})_a$ *and* $t = a + nh$, $n \in \mathbb{N}_0$. *Then, the* $\mathcal{Z}$-*transform of the Grunwald–Letnikov* h-*difference operator is*

$$\mathcal{Z}\left(\Delta^{\mu}_{h}y\right)(z) = z\left(\frac{hz}{z-1}\right)^{-\mu}\mathcal{Z}[y(a+nh)](z).$$

$$(1.78)$$

1.5.2. *Laplace Transform Method*

The Laplace transform method is proved to be an indispensable tool for solving engineering problems. We briefly introduce the Laplace transform method on the time scale $\mathbb{N}_a$ for fractional order difference equations. We will start with the Laplace transform of the fractional difference summation. We recall that the Laplace transform of a function $g : \mathbb{N}_a \to \mathbb{R}$ on an unbounded time scale $\mathbb{T}_a$ is [19]

$$\left(\mathcal{L}_a g\right)(s) = \int_a^{\infty} \exp^{\sigma}_{\ominus}(t,a)g(t)\Delta t, \quad \forall s \in \mathcal{D}\{g\},$$

$$(1.79)$$

where $\mathcal{D}\{g\}$ is the set for all complex constants for which the integral converges.

On the time scale $\mathbb{N}_a$, the Laplace transform of a function $g : \mathbb{N}_a \to \mathbb{R}$ is given as, [20]:

$$\left(\mathcal{L}_a g\right)(s) = \int_0^{\infty} \frac{g(t)}{(s+1)^{t-a+1}}\Delta t = \int_0^{\infty} \frac{g(t+a)}{(s+1)^{t+1}}\Delta t$$

$$= \sum_{k=0}^{\infty} \frac{g(k+a)}{(s+1)^{k+1}},$$

$$(1.80)$$

for all complex numbers $s \neq -1$ such that this improper integral (infinite series) converges. If $g : \mathbb{N}_a \to \mathbb{R}$ is of exponential order $r > 0$, i.e. if there exists a constant $M > 0$ such that

$$|g(t)| \leq Mr^t, \quad [\forall t \in \mathbb{N}_a]. \tag{1.81}$$

Then, the Laplace transform of g, $(\mathcal{L}_a g)(s)$, exists for all $s \in \mathbb{C} \backslash \overline{B_{-1}(r)}$.

The following lemma describes the exponential order of function g that relates to the exponential orders $\left(\Delta_a^{-\mu} g \right)(t)$ and $\left({}^{RL}\Delta^{\mu} g \right)(t)$.

Lemma 3. [20] *Suppose that* $g : \mathbb{N}_a \to \mathbb{R}$ *is of exponential order* $r \geq 1$ *and let* $\mu > 0$ *with* $n - 1 < \mu \leq n$. *Then for each fixed* $\epsilon > 0$, $\Delta_a^{-\mu} g$ *and* $\Delta_a^{\mu} g$ *are of exponential order* $r + \epsilon$, *and both* $\mathcal{L}_{a+\mu-n}\{\Delta_a^{-\mu} g\}(s)$ *and* $\mathcal{L}_{a+\mu-n}\{{}^{RL}\Delta^{\mu} g\}(s)$ *converge for all* $s \in \mathbb{C} \backslash \overline{B_{-1}(r)}$.

Let us now evaluate the Laplace transform of the fractional difference summation $\left(\mathcal{L}_{a+\mu-n}\{\Delta_a^{-\mu} f\} \right)(s)$.

Theorem 1.13 ([6]). *Suppose* $g : \mathbb{N}_a \to \mathbb{R}$ *is of exponential order* $r \geq 1$ *and let* $\mu > 0$ *with* $n - 1 < \mu \leq n$. *Then,* $\left(\mathcal{L}_{a+\mu-n}\{\Delta_a^{-\mu} f\} \right)(s)$ *converges for all* $s \in \mathbb{C} \backslash \overline{B_{-1}(r)}$, *and*

$$\mathcal{L}_{a+\mu-n}\{\Delta_a^{-\mu} g\}(s) = \frac{(s+1)^{\mu-n}}{s^{\mu}} \mathcal{L}_a\{g\}(s). \tag{1.82}$$

Now let us proceed to the evaluation of the Laplace transform of the Riemann–Liouville fractional difference operator.

Theorem 1.14. [6] *Let* $g : \mathbb{N}_a \to \mathbb{R}$ *with exponential order* $r \geq 1$ *and let* $\mu > 0$ *with* $n - 1 < \mu \leq n$. *Then for all* $s \in \mathbb{C} \backslash \overline{B_{-1}(r)}$, *we have*

$$\mathcal{L}_{a+\mu-n}\{{}^{RL}\Delta_a^{\mu} g\}(s) = s^{\mu}(s+1)^{n-\mu} \mathcal{L}_a\{g\}(s)$$

$$- \sum_{j=0}^{n-1} s^j \Delta_a^{\mu-1-j} g(a+n-\mu). \tag{1.83}$$

The third type of the operator, which is taken into our consideration, is the fractional left-Caputo difference operator.

Theorem 1.15. *Let* $g : \mathbb{N}_a \to \mathbb{R}$ *with exponential order* $r \geq 1$ *and let* $\mu > 0$ *with* $n - 1 < \mu \leq n$. *Then for all* $s \in \mathbb{C} \backslash \overline{B_{-1}(r)}$, *we have*

$$\mathcal{L}_{a+\mu-n}\{{}^c\Delta_a^\mu g\}(s) = s^\mu (s+1)^{n-\mu} - \sum_{j=0}^{n-1} s^j \Delta_a^{\mu-1-j} g(a+n-\mu).$$

(1.84)

1.6. Stability of Fractional Order Difference Systems

The asymptotic stability of fractional order difference equations is an important topic. This serves as the basis for demonstrating that the system state converges to zero as $t \to \infty$. This is especially important for stabilizing and synchronizing fractional order discrete-time dynamical systems. Consider the left-Caputo fractional difference system

$$^C\Delta_a^\mu x(t) = g(t+\mu-1, x(t+\mu-1)), \quad t \in \mathbb{N}_{a+1-\mu}, \qquad (1.85)$$

with initial condition $x(a)$, fractional order $\mu \in (0,1]$ and $f : \mathbb{N}_a \times \mathbb{R}^n \to \mathbb{R}^n$ being continuous. The constant vector x_f is an equilibrium point if and only if $g(t, x_f) = 0$ for all $t \in \mathbb{N}_a$. Without loss of generality, let the equilibrium point be $x_f = 0$. If the equilibrium point $x_f \neq 0$, it can be shifted to the origin by the change of variable $y = x - x_f$.

After defining the equilibrium point, the generalized stability of the fractional order discrete system will be discussed in the next subsection.

1.6.1. *Stability of Fractional Order Linear Systems*

First of all, we consider the integer order linear difference system

$$\begin{cases} \Delta z(m) = B g\left(z(m)\right), & m \in \mathbb{N} \\ z(0) = z_0, \end{cases} \qquad (1.86)$$

where $B \in \mathcal{M}_n(\mathbb{Z}^+)$ and $z(m) \in \mathbb{R}^n$.

Theorem 1.16. [17] *The solution of system* (1.86) *is:*

- *Globally asymptotically stable, if all the eigenvalues* λ_i, $(i = 1, \ldots, n)$ *of* B *satisfy* $|\lambda_i + 1| < 1$.
- *Unstable, if there is an eigenvalue* λ_i *of* B *such that* $|\lambda_i + 1| > 1$.

Similarly, we assume the following Caputo fractional order linear difference equation

$$^C\Delta_a^\mu y(n + 1 - \mu) = Ay(n), \tag{1.87}$$

where $x(t) \in \mathbb{R}^{n \times n}$, $A \in \mathbb{R}^{n \times n}$ and $0 < \mu < 1$ is the fractional order. Then we have the following result, which was introduced in [22].

Theorem 1.17. *Let* $\mu \in (0, 1]$ *and* $A \in \mathbb{R}^{n \times n}$. *Then* (1.87) *is asymptotically stable if and only if the isolated zeros, off the non-negative real axis, of* $\det(1 - z^{-1}(1 - z^{-1})^{-\mu}A)$ *lie inside the unit circle.*

Remark 9. If $\mu \to 1^-$, then Theorem 1.17 simplifies to the isolated zeroes, the *non-negative real axis*, of

$$\det\left(I - \frac{1}{z-1}A\right) = \frac{1}{z-1}\det((z-1)I - A), \tag{1.88}$$

lie inside the unit disk. This means that all the eigenvalues $\lambda = (z-1)$ of A satisfy $|(z-1) + 1| = |z| < 1$.

Based on the above theorem, the following lemma was recently derived.

Lemma 4. *If* $\mu = \frac{1}{2}$, *then the zero solution of Eq.* (1.87) *is asymptotically stable if and only if* $-\sqrt{2} < \lambda < 0$.

Example 7. Consider the following $\frac{1}{2}$-order difference system

$$\begin{cases} \Delta^{\frac{1}{2}}_{a-\frac{1}{2}} y(t) = \lambda y\left(t - \frac{1}{2}\right), & t \in \mathbb{N}_a, \text{ and } \lambda \in \mathbb{R}, \\ y\left(a - \frac{1}{2}\right) = 1. \end{cases} \tag{1.89}$$

Recall that Corollary 4 asserts that the zero solution of Eq. (1.89) is asymptotically stable if $-\sqrt{2} < \lambda < 0$. This, obviously, is evident in Figure 1.1, which depicts the solution $\{yn\}$ for different values of λ.

A similar kind of proof can be extended for this theorem.

Theorem 1.18. [23] *Let $x = 0$ be an equilibrium point for the fractional order linear difference system (1.87) where $\mu \in (0,1]$ and $A \in \mathbb{R}^{n \times n}$. Then, system (1.87) is asymptotically stable if and only if the isolated zeros, off the non-negative real axis, of $det(1 - z^{-1}(1 - z^{-1})^{-\mu}A)$ lie inside the unit circle.*

Example 8. Consider the following $\frac{1}{2}$-order systems of difference equations

$$\Delta^{\frac{1}{2}}_{a-\frac{1}{2}} y(t) = \begin{bmatrix} \lambda_1 & 1 \\ 0 & \lambda_2 \end{bmatrix} y\left(t - \frac{1}{2}\right), \quad t \in \mathbb{N}_a, \ a \in \mathbb{R}, \text{ and } \lambda_1, \lambda_2 \in \mathbb{R}, \tag{1.90}$$

subject to the initial condition

$$y\left(a - \frac{1}{2}\right) = \begin{bmatrix} 1 \\ 1 \end{bmatrix}.$$

Recall that Corollary 4 asserts that the zero solution of (1.90) is asymptotically stable if $-\sqrt{2} < \lambda_1, \lambda_2 < 0$. For different values of λ_i, it is confirmed in Figure 1.2 below. The figure depicts the 2-norm of the solution $\{y_n\}$ for the first 100 iterates.

One more stability theorem for $0 < \mu < 1$ was proposed by Cermak *et al.* in [22] based on an extension of the stability theorem of the integer order linear difference equations, and by analyzing the geometric property of the stability domain.

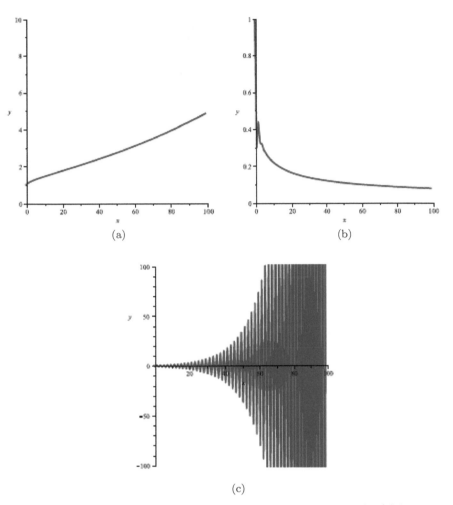

Figure 1.1. The first 100 iterates of the solution yn of Example 7, for (a) $\lambda = 0.1$, (b) $\lambda = -0.7$, and (c) $\lambda = -1.5$.

Theorem 1.19. [22] *The trivial solution of the fractional order linear difference system* $^{C}\Delta_a^\mu x(t) = g(t + \mu - 1, x(t + \mu - 1))$ *where* $\mu \in (0, 1)$ *is asymptotically stable, if*

$$\lambda \in= S^\mu \left\{ z \in \mathbb{C} : \ |z| < \left(2 \cos \frac{|\arg z| - \pi}{2 - \mu} \right)^\mu \ and \ |\arg z| > \frac{\mu \pi}{2} \right\},$$

$$(1.91)$$

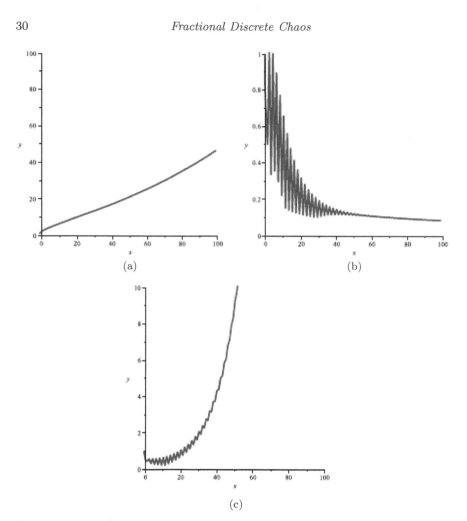

Figure 1.2. The first 100 iterates of 2-norm of the solution y_n of Example 8, for (a) $\lambda_1 = 0.1$, $\lambda_2 = 0.01$, (b) $\lambda_1 = -1.3$, $\lambda_2 = -1$, and (c) $\lambda_1 = -1.5$, $\lambda_2 = -0.7$.

for all eigenvalues λ of A. In this case, the solutions of (1.87) decay towards zero algebraically (and not exponentially), more precisely

$$\|y(n)\| = O\left(n^{-\mu}\right) \quad \text{as } n \to \infty, \tag{1.92}$$

for any solution x of (1.87). Furthermore, if $\lambda \in \mathbb{C}(S^\mu)$ for an eigenvalue λ of A, the zero solution of (1.87) is not stable.

Remark 10. The assertions of Theorems 1.19 and 1.18 describe the same stability region, but these analytical descriptions are different.

In particular, the condition stated in Theorem 1.19 seems to be more convenient for practical purposes due to the explicit form of S^μ. Also, it enables us to collect the following basic properties of S^μ.

Example 9. Consider the following fractional linear difference system

$$\begin{cases} {}^C\Delta_a^\mu x(t) = -x(t-1+\mu) + y(t-1+\mu), \\ {}^C\Delta_a^\mu y(t) = -y(t-1+\mu) + z(t-1+\mu), \\ {}^C\Delta_a^\mu z(t) = z(t-1+\mu), \end{cases} \quad (1.93)$$

where $a = 0$ and $t \in \mathbb{N}_{1+\mu}$, and the matrix A is given by

$$A = \begin{pmatrix} -1 & 1 & 0 \\ 0 & -1 & 1 \\ 0 & 0 & 1 \end{pmatrix}.$$

We can see that the characteristic equation of the matrix A is given by $(-1-\lambda)^2(1-\lambda) = 0$. One can observe that the eigenvalues λ_i of the matrix A satisfy

$$|\arg \lambda_i| = \pi \frac{\mu\pi}{2} \quad \text{and} \quad |\lambda_i| = 1 \left(2\cos \frac{|\arg \lambda_i| - \pi}{2 - \mu} \right).$$

Therefore, by Theorem 1.19 the trivial solution is asymptotically stable. The state evolution of the linear fractional system is depicted in Figure 1.3.

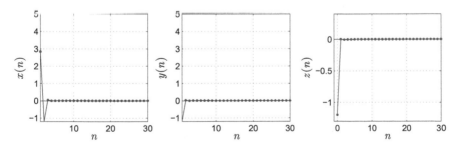

Figure 1.3. The evolution of the solution yn of Example 9, for $\mu = 0.98$.

1.6.2. *Stability of Fractional Order Nonlinear Difference Systems*

The Lyapunov direct method is known to provide a platform for analyzing system stability without solving the difference equations. This analysis is extended to the h-left-Caputo fractional difference systems.

Definition 13. [24] A function $\gamma(t)$ is said to belong to the class $\mathcal{K}$ if and only if $\gamma \in C[(0,\gamma), \mathbb{R}_+]$, $\gamma(0) = 0$, and $\gamma(t)$ is strictly monotonically increasing in t. Then we present the following theorem of a discrete fractional Lyapunov direct method for system (1.94).

Definition 14. A real-valued function $V(t,y)$ defined on $(h\mathbb{N})_a \times S_\gamma$, where $S_\gamma = \{y \in \mathbb{R}^n : ||y|| \leq \gamma$, is said to be decrescent if and only if $V(t,0) = 0$ for all $t \in (h\mathbb{N})_a$ and there exists $\gamma(t) \in \mathcal{K}$ such that $V(t,y) \leq \gamma(t)$, $||y|| = t$, $(t,x) \in (h\mathbb{N})_a \times S_\gamma$.

Theorem 1.20. [24] *Let $y = 0$ be an equilibrium point for the fractional order difference system* (1.94)

$$\begin{cases} {}^{C}_{h}\Delta^{\mu}_{a} y(t) = g(t + \mu h, y(t + \mu h)), \\ y(a) = y_0 \in \mathbb{R}^n, \quad [t \in (h\mathbb{N})_{a+(1-\mu)h}], \end{cases} \tag{1.94}$$

where g is continuous with respect to y, and $\mu \in (0,1]$. Assume that there exist a function $V(t,y)$, which is a positive definite and decrescent scalar discrete, such that

$$\gamma_1(||y(t)||) \leq V(t, y(t)) \leq \gamma_2(||y(t)||), \quad t \in (h\mathbb{N})_a \tag{1.95}$$

$${}^{C}_{h}\Delta^{\mu}_{a} V(t, y(t)) \leq -\gamma_3(||y(t+\mu h)||), \quad t \in (h\mathbb{N})_{a+(1-\mu)h}, \tag{1.96}$$

where $\mathcal{K}$ γ_1, γ_2 and γ_3 are arbitrary positive constant. Then, the fractional difference system (1.94) *is asymptotically stable.*

One should note that the construction of γ_i functions is not easy. In this section, we present another Lemma and sufficient condition. We start by presenting the following lemma.

Lemma 5 (*Discrete comparison principal*). *For* $\mu \in (0, 1]$, $_h^C\Delta_a^\mu y(t) \geq C_h\Delta_a^\mu y(t)$, $\forall t \in (h\mathbb{N})_{a+(1-\mu)h}$ *and* $x(a) = y(a)$. *Then*

$$x(t + \mu h) \geq y(t + \mu h).$$

An useful inequality for Lyapunov functions is now provided.

Lemma 6. *For any discrete time* $t \in (h\mathbb{N})_{a+(1-\mu)h}$, *the following inequality holds*

$$_h^C\Delta_a^\mu x^2(t) \leq 2x(t + \mu h)_h^C\Delta_a^\mu x(t), \quad 0 < \mu \leq 1. \tag{1.97}$$

Remark 11. For $\mathbf{y} = (y_1(t), \ldots, y_m(t))^T$, $t \in (h\mathbb{N})_{a+(1-\mu)h}$, Lemma 6 still holds. For example, we can have Lemma 6 as

$$_h^C\Delta_a^\mu(x^T(t)x(t)) \leq 2x^T(t + \mu h)_h^C\Delta_a^\mu x(t), \quad t \in (h\mathbb{N})_{a+(1-\mu)h}$$

Example 10. Consider the following nonlinear fractional difference system

$$\begin{cases} _h^C\Delta_a^\mu x_1(t) = -x_1(t + \mu h) + x_2^3(t + \mu h), \ 0 < \mu \leq 1, \\ _h^C\Delta_a^\mu x_2(t) = -x_1(t + \mu h) - x_2(t + \mu h), \ t \in (h\mathbb{N})_{a+(1-\mu)h}, \end{cases} \tag{1.98}$$

with the initial conditions $x_1(0) = 0.4$ and $x_2(0) = 0.8$.

We use the Lyapunov function $V = \frac{1}{2}x_1^2(t) + \frac{1}{4}x_2^4(t)$. According to Lemma 6, we have

$$_h^C\Delta_a^\mu V \leq x_1(t + \mu h)_h^C\Delta_a^\mu x_1(t) + \frac{1}{2}x_1^2(t + \mu)_h^C\Delta_a^\mu x_2^2(t) \tag{1.99}$$

$$\leq x_1(t + \mu h)_h^C\Delta_a^\mu x_1(t) + x_2^3(t + \mu)_h^C\Delta_a^\mu x_2(t) \tag{1.100}$$

$$= -x_1^2(t + \mu h) - x_2^4(t + \mu h) < 0. \tag{1.101}$$

As a result, the system is asymptotically stable from Theorem 1.20. Using the h-fractional sum operator reported in this chapter, we can derive the numerical formula of fractional nonlinear system (1.98).

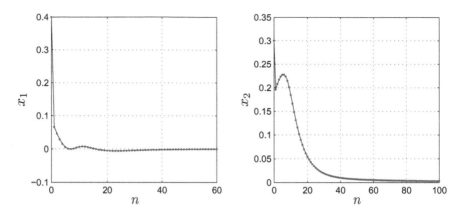

Figure 1.4. Asymptotic stability of states x_1 and x_2 for $\mu = 0.9$.

Note that, the numerical formula obtained is an implicit formula. To plot the states of the system, we use the Newton method to transfer it into an explicit one. The results are illustrated in Figure 1.4 which proves the theoretical results.

Chapter 2

Chaotic Methods and Tests

2.1. Introduction

Nonlinear systems are very interesting to engineers, physicists and mathematicians because most real physical systems are inherently nonlinear in nature [25]. In mathematics, a nonlinear system is any problem, where the variables to be solved cannot be written as a linear combination of independent components. If the equation contains a nonlinear function (power or cross product), the system is nonlinear as well. The system is nonlinear if there is some typical nonlinearity as, for instance, saturation, hysteresis, etc. These characteristics are basic properties of nonlinear systems.

Nonlinear equations are difficult to be solved by analytical methods and give rise to interesting phenomena such as bifurcation and chaos. Even simple nonlinear (or piecewise linear) dynamical systems can exhibit completely unpredictable behavior, the so-called deterministic chaos. Dynamical systems constitute a mathematical framework common to many disciplines, among which are ecology and population dynamics. Chaos theory has been so surprising because it can also be found within trivial systems. Discrete chaotic systems have been around for a while. Hence we are forced to resort to different means in order to understand these behaviors. Moreover, discrete-time systems can avoid the calculation error of numerical discretization of continuous ones.

Discrete-time dynamical systems can be written as the map [26]

$$\mathbf{x}(n+1) = f(\mathbf{x}(n)), \tag{2.1}$$

35

where $\mathbf{x}_n$ is N-dimensional, $\mathbf{x}(n) = (x_1(n), x_2(n), \ldots, x_N(n))$, which is a shorthand notation for

$$\begin{cases} x_1(n+1) = f_1(x_1(n), x_2(n), \ldots, x_N(n)), \\ \quad\vdots \\ x_N(n+1) = f_N(x_1(n), x_2(n), \ldots, x_N(n)), \end{cases} \quad\quad (2.2)$$

with n denoting the discrete-time variable. Given initial values $(x_1(0), \ldots, x_N(0))$, we can compute $(x_1(n), \ldots, x_N(n))$ successively for all positive n using (2.1). Thus, the trajectory of $\mathbf{x}(0)$ is the sequence

$$\mathbf{x}(0), \quad \mathbf{x}(1) = f(\mathbf{x}(0)), \quad \mathbf{x}(2) = f^2(\mathbf{x}(0)), \ldots, \mathbf{x}(n) = f^n(\mathbf{x}(0)), \ldots$$

2.2. Discrete Chaos

Chaos theory is a branch of mathematics that has focused on the study of the behavior of dynamical systems that are very sensitive to initial conditions. Sensitive dependence here means that even a small change in one state of a deterministic nonlinear system can make a big differences in later states. This behavior is known as deterministic chaos, or simply named chaos. There is no widely accepted definition of chaos. One of the most popular definition of chaos, is the one given by Devaney [30], in which the dynamical system must exhibit sensitive dependence to initial conditions, topological transitivity, and dense periodic orbits [31]. Later it was proven that if a system is transitive with dense periodic orbits, then obviously sensitivity dependence to initial condition is guaranteed.

2.2.1. *Characterization of Chaotic Dynamical System*

2.2.1.1. *Sensitive initial condition*

Being sensitive to initial conditions is a phenomena first discovered by Poincaré in XIXth century, and rediscovered by Lorenz during the study of meteorology in 1963. The sensitive dependence to initial condition, also known as the *butterfly effect*, describes how a small change in one state of a deterministic nonlinear system may lead to dramatic change in the behavior of the system over time. This makes prediction of future behavior impossible, but this does not mean the system is not deterministic. The mathematical definition is given as follows.

Definition 15. [27] Let $\mathbf{X}$ be a compact metric space and f a continuous map. A dynamical system $(\mathbf{X}, f)$ has sensitivity dependence on initial conditions if $\exists \delta > 0$ *such that, for* $x \in \mathbf{X}$ *and each* $\epsilon > 0$, *there is* $y \in \mathbf{X}$ *with* $d(x, y) < \epsilon$ *and* $n \in \mathbb{N}$ *such that* $d(f^n x, f^n y) > \delta$.

Numerically, sensitivity is measured by a *Lyapunov exponent* such that a positive value implies the system is really sensitive to initial conditions. We will discuss this point later.

2.2.2. *Strange Attractor*

The strange attractor is a geometric feature of chaos. There is no rigorous definition of strange attractor, and all the definitions found in the literature are restrictive. Ruelle in [28] tries to give a mathematical definition which is summarized as follows.

Definition 16. A bounded set $\mathbf{A}$ in N-dimensional space, with transformation $\mathbf{f}$, is a *strange attractor* if there is a set $\mathbf{U}$ with the following properties [29]:

- It is an attractor, i.e. $\mathbf{A}$ is contained in $\mathbf{U}$.
- The bounded set is attracting, i.e. for every initial point $\mathbf{x}(0)$ in $\mathbf{U}$, any orbit by transformation $\mathbf{f}$ remains in $\mathbf{U}$. Moreover, any such orbit becomes and stays as close as one wants to A.
- Orbits whose initial points in $\mathbb{R}$ are extremely sensitive to initial conditions.
- It cannot split into two different attractors, i.e. if we choose a point $\mathbf{x}(0)$ in $\mathbf{A}$ such that it is arbitrary close to another point in $\mathbf{A}$, there is a point $\mathbf{x}(n)$ for some positive n.

Remark 12. The property of sensitive dependence on initial conditions makes the attractor strange.

2.2.3. *Chaotic Discrete Systems*

In practice, discrete-time dynamical systems (maps) are accurate models for natural physical phenomena in the field of biology, chemistry and physics. In the following, we give some famous examples of discrete dynamical systems.

Logistic map

The logistic map can be obtained by the discretization of logistic differential equation that was initially proposed in the population growth model by Verhulst [32]. The discretized logistic equation or logistic map:

$$x(n + 1) = Kx(n)\left(1 - x(n)\right). \tag{2.3}$$

The logistic map exhibits a complex dynamic behavior when $K = 3.57$. When the latter value is reached, an attractor appears as seen in Figure 2.1.

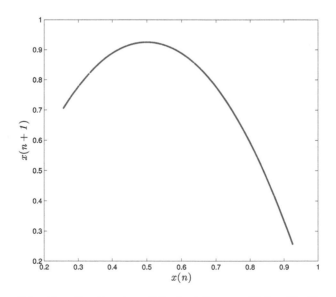

Figure 2.1. Chaotic attractor of the logistic map (2.6) with $K = 3.57$.

Hénon map

The Hénon map is a two-dimensional discrete-time system given by

$$\begin{cases} x_1(n+1) = 1 - \alpha x_1^2(n) + x_2(n), \\ x_2(n+1) = \beta x_1(n), \end{cases} \tag{2.4}$$

where α and $|\beta| \leq 1$ are external parameters. This map was first discussed by Hénon in 1976 [33] as the Poincaré map of the famous continuous-time Lorenz system, and it can be considered as an extension of the logistic map. The Hénon map has a strange attractor for the standard parameter values $\alpha = 1.4$ and $\beta = 0.3$, as shown in Figure 2.2.

Flow chaotic map

The Flow map is a two-dimensional chaotic discrete-time system, given by:

$$\begin{cases} x(n+1) = y(n) + \theta x(n), \\ y(n+1) = \lambda + x^2(n) - y(n), \end{cases} \tag{2.5}$$

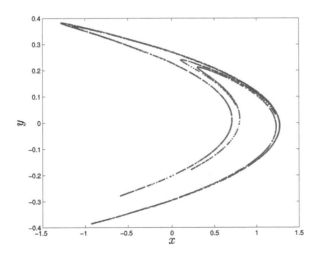

Figure 2.2. The chaotic attractor of the Hénon map (2.4).

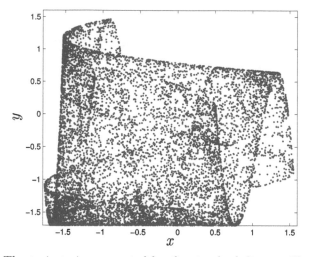

Figure 2.3. The trajectories generated by the standard discrete Flow map with $\theta = -0.1$, $\lambda = -1.7$, and $(x(0), y(0)) = (0, 1)$.

where θ and λ are the system's parameters. Figure 2.3 shows a chaotic attractor of the Flow map obtained for the system parameters $\theta = -0.1$ and $\lambda = -1.7$.

2.3. Classical Tools to Detect Fractional Chaos

The theory and applications of discrete fractional calculus may be considered as a novel topic, some recent contributions deal with discrete analogues of continuous fractional calculus and fractional difference equations. Very recently, some literature has been studied on the dynamics, including chaotic behavior of discrete-time fractional order systems using the Caputo fractional difference operator type, as introduced in Chapter 1. The first such system is the fractional order logistic map which was proposed by Wu and Baleanu [34] where chaotic behaviors and a synchronization method of the fractional map were numerically illustrated.

The literature seems to agree that fractional chaotic maps possess superior properties as compared to their standard counterpart. For instance, the general dynamics of fractional maps are heavily dependent on variations in the fractional order [34]. This adds new degrees of freedom to the map's states making them more suitable for data encryption, for example. In addition, Edelman [35] showed that the convergence speed as well as the convergence route depend on the initial conditions, which leads to richer dynamics. The fractional difference provides us a new powerful tool to characterize the dynamics of discrete complex systems more deeply. Recently, Peng *et al.* revised the fractional logistic map reported by Wu *et al.*, based on Edelman's work and presented the correct simulation results for bifurcation diagrams in [36]. So we start by reporting both fractional Logistic maps to clarify the method used here.

The fractional Logistic map

Consider the logistic map defined above

$$x(n+1) = Kx(n)\left(1 - x(n)\right), \quad n = 1, 2, \dots. \tag{2.6}$$

First, we take the first order difference equation of (2.6), as

$$\Delta x(n) = Kx(n)\left(1 - x(n)\right) - x(n), \quad n = 1, 2, \dots. \tag{2.7}$$

We may replace the standard difference in (2.7) with the Caputo-difference operator defined in Chapter 1 which yields

$$^C\Delta_a^\nu x(t) = Kx(t - 1 + \nu)\left(1 - x(t - 1 + \nu)\right) - x(t - 1 + \nu),$$

$$\text{for } t \in \mathbb{N}_{a+1-\nu} \tag{2.8}$$

where a is the starting point. The case $\mu = 1$ corresponds to the nonfractional scenario. To investigate the dynamics of the logistic map (2.8), we will need a discrete numerical formula that allows us to evaluate the states of the map in fractional discrete time. According to Wu *et al.* and other similar studies, we can obtain the following equivalent discrete integral from Eq. (2.8) by using Theorem 1.11 as reported in the first chapter

$$x(t) = x(a) + \frac{1}{\Gamma(\nu)} \sum_{s=1-\nu}^{t-\nu} (t - s - 1)^{(\nu-1)}$$

$$\times \left(Kx(s - 1 + \nu)\left(1 - x(s - 1 + \nu)\right) - x(s - 1 + \nu)\right), \tag{2.9}$$

for $t \in \mathbb{N}_1$. As a result, the numerical formula can be presented accordingly

$$x(n) = x(0) + \frac{1}{\Gamma(\nu)} \sum_{j=1}^{n-1} \frac{\Gamma(n-1-j+\nu)}{\Gamma(n-j)} (Kx(j)(1-x(j)) - x(j)),$$

(2.10)

where $x(0)$ is the initial condition. Compared with the map of the integer order, the fractionalized one (2.8) has a discrete kernel function. $x(n)$ depends on the past information $x(0), \ldots, x(n-1)$. As a result, the memory effects of the discrete maps mean that their present state of evolution depends on all past states.

In order to be able to carry out the analysis and numerical simulations required for discrete fractional system, we will introduce the basic mathematical concepts and numerical tools to analyze such irregular geometrical entities. The numerical tools introduced herein are bifurcation diagrams, Lyapunov exponents, 0-1 test, C_0 complexity, and approximate entropy.

2.3.1. *Bifurcation Diagrams*

With the discovery of chaotic dynamics, the theory has become even more important, as researchers are trying to find mechanisms by which systems change from simple to highly complicated behavior. Consider an nth-order discrete-time system

$$x(n+1) = f(x(n), \alpha),$$

(2.11)

with a parameter $\alpha \in \mathbb{R}$. As α changes, the dynamic behavior of the system is also changed. Typically, a small change in α produces small quantitative changes in the states of the system. Such change in the specific behavior of the map is known as a *bifurcation.* Accordingly, the bifurcation phenomenon describes the fundamental alteration in the dynamics of nonlinear systems under parameter variation. For this reason, it is considered as a tool that helps to understand equilibrium loss and its consequences for complex behavior. Moreover, bifurcation diagrams show some characteristic properties of asymptotic solutions of a dynamical system as a function of a control parameter, allowing one to immediately where qualitative changes in the asymptotic solutions occur. The parameter at which the dynamic behavior changes is called the bifurcation parameter.

2.3.2. *Lyapunov Exponents*

Perhaps, the most important qualitative measure of chaos is the method of Lyapunov exponents, which was originally developed by the Russian mathematician Aleksander Lyapunov in his research paper [37]. The Lyapunov exponent (LE) is a numerical characteristic that measures the average divergence or convergence of a small perturbation introduced in the orbit of a dynamical system, and it is one of the numerical characteristics used to determine chaotic behavior [38, 39]. The LE calculation methods start with exponential divergence of nearby trajectories when the trajectory is on the attractor. To see this, consider a trajectory $\mathbf{x}(n)$ starting from an initial condition $\mathbf{x}(0)$ and a trajectory $\mathbf{x}'$ stemming from a point $\mathbf{x}'(0)$ very close to $\mathbf{x}(0)$. Let $\delta\mathbf{x}(n)$ be the distance after n iterations, i.e. $\delta(n) = ||\mathbf{x}'(n) - \mathbf{x}(n)||$. When there is sensitivity to initial conditions, there is an exponential divergence between the two trajectories (see Figure 2.4). The exponential separation between the two trajectories is given by

$$\delta(n) \approx \delta(0)e^{n\lambda}, \tag{2.12}$$

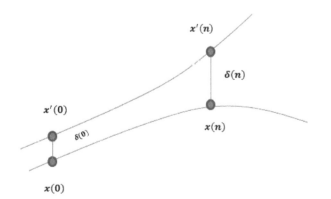

Figure 2.4. The divergence of distance for two nearby orbits.

where λ is the Lyapunov exponent and $\delta(0)$ is the initial trajectory separation. There are, in fact, several Lyapunov exponents. One exponent is defined for each dimension, representing the average rate of growth or decay along each of the principal axes in the d_E-dimensional state space. But for a long time, the behavior of $\delta(n)$ is dominated by the maximum Lyapunov exponents.

Measuring the Lyapunov exponent is always an important problem no matter in a fractional order system or in an integer order system. When the discrete dynamical system is available, the Jacobian method is well applied. In this section, we consider the extended Jacobian matrix algorithm which was proposed in [46] by Wu and Balaneau. Several well known fractional order maps are presented as examples to estimate the LEs, and to validate the efficiency of the proposed method. The concept of Lyapunov exponents (LEs) in integer order discrete-time systems is first introduced.

2.3.2.1. *Calculating Lyapunov exponents via Jacobi method*

Let us consider N-dimensional discrete dynamical system, defined by the following difference equations

$$\mathbf{x}(n+1) = \mathbf{F}\left(\mathbf{x}(n)\right), \qquad (2.13)$$

where $\mathbf{x}(n) = \left(\mathbf{x}^{(1)}(n), \ldots, \mathbf{x}^{(N)}(n)\right)^T$ and $\mathbf{F}$ is the corresponding vector field. To determine the N-Lyapunov exponents of the system, one should consider a small N-dimensional sphere of initial conditions in the phase space of the discrete dynamical system (2.13). An example of the change in the sphere of initial conditions is given in Figure 2.5. After few iterations n, this sphere transforms into an ellipsoid whose principal axes stretch or contract at rates given by the LEs. This is determined by the $N \times N$ Jacobian matrix of (2.13), which can be given as

$$J_n = D_x \mathbf{F}^n\left(\mathbf{x}(0)\right) = \begin{bmatrix} \frac{\partial \mathbf{F}_1^n}{\partial x_1} & \cdots & \frac{\partial \mathbf{F}_1^n}{\partial x_N} \\ \cdots & \cdots & \cdots \\ \frac{\partial \mathbf{F}_N^n}{\partial x_1} & \cdots & \frac{\partial \mathbf{F}_N^n}{\partial x_N} \end{bmatrix} \left(\mathbf{x}^{(1)}(0) \ldots \mathbf{x}^{(N)}(0)\right),$$

$$(2.14)$$

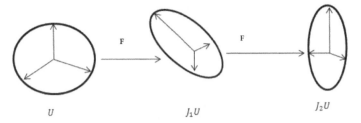

Figure 2.5. This figure illustrates change in a sphere of initial conditions under a map.

in which $D_x\mathbf{F}^n(\mathbf{x}(0)) = J\left(\mathbf{F}^{n-1}(\mathbf{x}(0))\right) \times J\left(\mathbf{F}^{n-2}(\mathbf{x}(0))\right)\ldots$ $J\left(\mathbf{F}(\mathbf{x}(0))\right) \times J(\mathbf{x}(0))$.

Consequently, according to the theorem of Oseledec (1968) in [38] we can define the average rate of growth as

$$\lambda_1 = \lim_{N \to \infty} \frac{1}{N} \ln \lVert J^n(x(0))\, u \rVert. \tag{2.15}$$

where the limit exists for almost all $x(0)$, and for almost all tangent vectors u, it is equal to the maximum Lyapunov exponent λ_1. It is sufficient to look at the maximum LEs for assessing chaotic dynamical systems. In particular, if $\lambda_1 > 0$ the dynamics will present, at least, one unstable direction in the phase space, that implies chaotic behavior.

When we write down the eigenvalue of J^n as $E_1(n,x) \geq \cdots \geq E_N(n,x)$, then, the ith Lyapunov exponents are defined as

$$\lambda_i = \lim_{n \to \infty} \frac{1}{n} \ln |F_i(n,x)|, \quad i = 1,\ldots,m. \tag{2.16}$$

The calculation of Lyapunov exponents using the expression (2.16) is extremely difficult because of the product $J\left(\mathbf{F}^{n-1}(\mathbf{x}(0))\right) \times J\left(\mathbf{F}^{n-2}(\mathbf{x}(0))\right)\ldots J\left(\mathbf{F}(\mathbf{x}(0))\right) \times J(\mathbf{x}(0))$. Even for few iterations, the components of matrix J^n become very large for chaotic attractors and null for the periodic attractors. To circumvent these problems, several algorithms have been proposed to calculate the LEs for many integer order discrete maps [41]. But if a system under consideration is with fractional order, the calculation of Lyapunov exponents is not straightforward. In recent years, Wu *et al.* [46] considered the extension of the Jacobian matrix algorithm to discrete fractional maps. We sketch the algorithm and numerical computation in the remainder of the next subsection with more details.

2.3.2.2. *Calculating Lyapunov exponents of discrete fractional map via Jacobian matrix algorithm*

Given an N-dimensional dynamical discrete system with fractional order and initial condition:

$$
\begin{cases}
{}^{C}\Delta_a^{\mu}\mathbf{x}(t) = g\left[\mathbf{x}(t + \mu - 1)\right], \\
\Delta^k(a) = \mathbf{x}(k), \quad N = \lceil \mu \rceil + 1, \quad k = 0, 1, \dots, N - 1,
\end{cases}
\tag{2.17}
$$

where ${}^{C}\Delta_a^{\mu}\mathbf{x}(t)$ is the left-Caputo operator with $\mu \in (0, 1]$ and $\mathbf{x}(t) = \{x_1(t), x_2(t), \dots, x_N(t)\}$ is the state vector.

As reported by Wu and Balaneau [46], the Jacobian matrix $J(n)$ of the fractional order map (2.17) is defined as

$$
J(n) = \begin{bmatrix}
a_{11}(n), & a_{12}(n), & \cdots, & a_{1m}(n) \\
\vdots & \vdots & \vdots & \vdots \\
a_{N1}(n), & a_{N2}(n), & \cdots, & a_{Nm}(n)
\end{bmatrix},
\tag{2.18}
$$

where $J(0) = I$ is the identity matrix. Note that the elements a_{ij} can be obtained by multiplying the tangent map of the right side of system (2.17) with the matrix J, as follows

$$
J(n) = \begin{bmatrix}
\frac{\partial g_1}{\partial x_1}, & \frac{\partial g_1}{\partial x_2}, & \cdots, & \frac{\partial g_1}{\partial x_m} \\
\vdots & \vdots & \vdots & \vdots \\
\frac{\partial g_m}{\partial x_1}, & \frac{\partial g_m}{\partial x_2}, & \cdots, & \frac{\partial g_m}{\partial x_m}
\end{bmatrix}
$$

$$
\times \begin{bmatrix}
a_{11}(n - 1), & a_{12}(n - 1), & \cdots, & a_{1N}(n - 1) \\
\vdots & \vdots & \vdots & \vdots \\
a_{N1}(n - 1), & a_{N2}(n - 1), & \cdots, & a_{NN}(n - 1)
\end{bmatrix}.
\tag{2.19}
$$

This will lead to the following tangent discrete system which holds the memory effect

$$
\left\{
\begin{aligned}
a_{11}(n) &= a_{11}(0) + \frac{1}{\Gamma(\mu)} \sum_{i=0}^{n-1} \frac{\Gamma(n-1-i+\mu)}{\Gamma(n-i)} a_{11}(i) \frac{\partial g_1}{\partial x_1}(i) + a_{21}(i) \frac{\partial g_1}{\partial x_2}(i) + \dots \\
&\quad + a_{N1}(i) \frac{\partial g_1}{\partial x_N}(i), \\[2ex]
a_{12}(n) &= a_{12}(0) + \frac{1}{\Gamma(\mu)} \sum_{i=0}^{n-1} \frac{\Gamma(n-1-i+\mu)}{\Gamma(n-i)} a_{12}(i) \frac{\partial g_1}{\partial x_1}(i) + a_{22}(i) \frac{\partial g_1}{\partial x_2}(i) + \dots \\
&\quad + a_{N2}(i) \frac{\partial g_1}{\partial x_N}(i), \\[2ex]
\vdots \quad &\qquad \vdots \qquad\qquad\qquad \vdots \ \dots\dots\dots \ \vdots \qquad\qquad \vdots \\[2ex]
a_{NN}(n) &= a_{N1}(0) + \frac{1}{\Gamma(\mu)} \sum_{i=0}^{n-1} \frac{\Gamma(n-1-i+\mu)}{\Gamma(n-i)} a_{N1}(i) \frac{\partial g_m}{\partial x_1}(i) + a_{N2}(i) \frac{\partial g_N}{\partial x_2}(i) + \dots \\
&\quad + a_{NN}(i) \frac{\partial g_N}{\partial x_N}(i),
\end{aligned}
\right.
\tag{2.20}
$$

After N iterations, the Lyapunov exponents are given by

$$
\lambda_i = \lim_{N \to \infty} \frac{1}{N} \ln |E_i^{(N)}|, \quad i = \overline{1, N},
\tag{2.21}
$$

where E_i are the eigenvalues of the matrix J.

In the following, the fractional Duffing map is used to show the effectiveness of the proposed method, and all experiments are performed using Matlab.

2.3.2.3. *Application in fractional Duffing map*

One of the maps that has been getting a lot of attention lately is the Duffing map, which is defined based on the Duffing oscillator [48, 49]. This map was first introduced in [50] to describe the transformation required by a certain laser ray represented by its discrete-time distance after one round trip inside the laser cavity. The Duffing map is presented as follows

$$
\left\{
\begin{aligned}
x(n+1) &= y(n), \\
y(n+1) &= -\beta x(n) + \alpha y(n) - y^3(n),
\end{aligned}
\right.
\tag{2.22}
$$

where α and β are bifurcation parameters, and in fractional order case as [51]

$$\begin{cases} {}^{C}\Delta_a^\mu x(t) = y(t-1+\mu) - x(t-1+\mu), \\ {}^{C}\Delta_a^\mu y(t) = -\beta x(t-1+\mu) + (\alpha-1)y(t-1+\mu) - y^3(t-1+\mu), \end{cases}$$
(2.23)

where ${}^{C}\Delta_a^\mu$ is Caputo-like difference operator with $\mu \in (0,1]$ and $t \in \mathbb{N}_{a+1-\mu}$. For numerical analysis purposes, the numerical formula is defined as

$$\begin{cases} x(n) = x(0) + \dfrac{1}{\Gamma(\mu)} \sum_{j=1}^{n-1} \dfrac{\Gamma(n-1-j+\mu)}{\Gamma(n-j)} [y(j) - x(j)], \\[4mm] y(n) = y(0) + \dfrac{1}{\Gamma(\mu)} \sum_{j=1}^{n-1} \dfrac{\Gamma(n-1-j+\mu)}{\Gamma(n-j)} \\[2mm] \qquad\qquad \times \left[-\beta x(j) + (\alpha-1)y(j) - y^3(j)\right]. \end{cases}$$
(2.24)

Particularly, the case of $\mu = 1$ can be reduced to the standard fractional Duffing map (2.22) where the integer order map that produced the chaotic attractor, given the initial conditions $[x(0), y(0)] = [0.3, 0.1]$ and system parameters $[\alpha, \beta] = [2.77, 0.2]$ is shown in Figure 2.6.

The Jacobian matrix of the fractional Duffing map is given as

$$J_i = \begin{pmatrix} a_i & b_i \\ c_i & d_i \end{pmatrix},$$
(2.25)

where

$$a_i = a_0 + \frac{1}{\Gamma(\mu)} \sum_{j=1}^{i-1} \frac{\Gamma(i-1-j+\mu)}{\Gamma(i-j)} [-a_i + c_i],$$

$$b_i = b_0 + \frac{1}{\Gamma(\mu)} \sum_{j=1}^{i-1} \frac{\Gamma(i-1-j+\mu)}{\Gamma(i-j)} [-b_i + d_i],$$

$$c_i = c_0 + \frac{1}{\Gamma(\mu)} \sum_{j=1}^{i-1} \frac{\Gamma(i-1-j+\mu)}{\Gamma(i-j)} \left[-\beta a_i + (\alpha - 3y_i^2 - 1) c_i\right],$$

$$d_i = d_0 + \frac{1}{\Gamma(\mu)} \sum_{j=1}^{i-1} \frac{\Gamma(i-1-j+\mu)}{\Gamma(i-j)} \left[-\beta b_i + (\alpha - 3y_i^2 - 1) d_i\right].$$

In the numerical experiments, we can select the initial condition $[x(0), y(0)] = [0.3, 0.1]$, and the length of the series of data as $n = 12000$. To verify the effectiveness of the method, we compute the

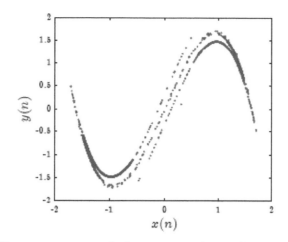

Figure 2.6. Chaotic attractor of the integer-order Duffing map (2.22) with $(\alpha, \beta) = (2.77, 0.2)$ and initial conditions $[x(0), y(0)] = [0.3, 0.1]$.

maximum Lyapunov exponent with the corresponding bifurcation diagrams. Figure 2.7 illustrates the bifurcation plots of (2.23) and the estimation of the maximum Lyapunov exponents by the Jacobian method with $\beta \in [-0.2, 1]$ and $\alpha = 2.77$. By direct comparison, the maximum LEs are consistent with the corresponding bifurcation diagrams. In the same way, it can be observed from Figure 2.7 that complex dynamics are coined in the fractional Duffing map, reflecting the dynamical effect of the fractional order in the proposed map. As stated in Figure 2.7, the maximum Lyapunov exponent is positive when $\beta \in]-0.1, 0.2]$ while it is negative in the rest of the interval which confirms the chaotic behavior of the fractional Duffing map.

2.4. 0–1 Test Method

The 0–1 test is an effective way to reflect on the sensitivity of the fractional order systems. This test was proposed by Gootwald GA and Melbourne I in [52] to identify the different behavior patterns of dynamical systems. In contrast to the Lyapunov exponents method, the 0–1 test is directly applied to the series, based on the statistical properties of a single coordinate of the dynamical system [53]. In addition, the dimension of the dynamical system and the form of

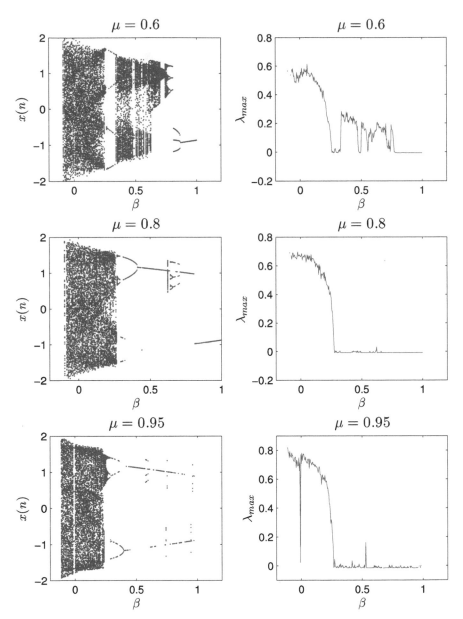

Figure 2.7. Bifurcation diagrams versus β of the fractional Duffing map and the corresponding maximum LEs diagrams.

the underlying equations are irrelevant [54], and the chaos of the series can be identified without reconstructing the phase space. Moreover, the outlined test is suitable for discrete as well as for continuous dynamical systems. Also, the applicability of the test to the analysis of fractional order continuous system was discussed in [55]. Similarly, the test was successfully applied to discrete systems with fractional order [60]. In this section, we briefly review how the test is implemented for fractional order discrete-time systems. Consider a set of discrete data $y(m)$, $m = 1, \ldots, N$ obtained from a fractional order discrete-time system. One can define the two variables, as

$$
p(m, \bar{c}) = \sum_{i=1}^{m} y_i \cos(i\bar{c}), \quad q(m, \bar{c}) = \sum_{i=1}^{m} y_i \sin(i\bar{c}), \tag{2.26}
$$

where $\bar{c} \in (0, 2\pi)$ is a constant. In order to study the boundedness or unboundedness of the new variables $p(m, \bar{c})$ and $q(m, \bar{c})$, we calculate the mean square displacement $M(m, \bar{c})$ of $p(m, \bar{c})$ and $q(m, \bar{c})$ as:

$$
M(m, \bar{c}) = \lim_{N \to +\infty} \left\{ \frac{1}{N} \sum_{j=1}^{N} [p(j + m) - p(i)]^2 + [q(j + m) - q(i)]^2 \right\}. \tag{2.27}
$$

The convergence and divergence of the new variables $p(m, \bar{c})$ and $q(m, \bar{c})$ can be measured by $M(m, \bar{c})$. Now, the correlation coefficient is used to estimate the asymptotic growth rate K_c; as

$$
K_{\bar{c}} = \frac{\operatorname{cov}(\delta, \zeta)}{\sqrt{\operatorname{var}(\delta)\operatorname{var}(\zeta)}} \in [-1, 1], \tag{2.28}
$$

where δ is the vector from 1 up to N, and ζ is the vector formed by the mean square displacement M_c.

As proposed by Gottwald and Melbroune [53, 54], the parameter $\bar{c}$ strongly influences the dynamics in the $p(\bar{c}) - q(\bar{c})$ plane, and give rise to spurious values for $K_{\bar{c}}$. The proposed solution for this issue is to calculate the value of the asymptotic growth rate K by taking the median of the correlation coefficient $K_{\bar{c}}$ for 100 values or more of $\bar{c} \in [0, \pi]$, i.e.

$$K = \text{median}(K_{\bar{c}}). \tag{2.29}$$

What is left to say is how to choose the parameters N and m. In the selection of N, it should tend to infinity in theory, but this is not feasible, in practice. Therefore, in practice the sequence length N should be selected long enough. For the selection of m, this was discussed by Gottwald and Melbroune [52] where they pointed out that the value of m is chosen based on the principle of $m \ll N$.

According to the 0–1 test method, the fractional discrete-time system shows chaotic behavior when K tends to 1 and the $p - q$ plane presents a Brownian-like trajectory, while when K approaches 0 and the $p-q$ graph displays a bounded periodic ring, the fractional discrete-time system shows regular characteristics.

2.4.1. *Applications of the 0–1 Test on Fractional Order Maps*

Let us now begin by examining the results of the 0–1 test applied to the fractional order Duffing map. Consider the numerical formula of the fractional order Duffing map.

$$\begin{cases} x\left(n\right) = x\left(0\right) + \dfrac{1}{\Gamma\left(\mu\right)} \sum_{j=1}^{n-1} \dfrac{\Gamma\left(n-1-j+\mu\right)}{\Gamma\left(n-j\right)} \left[y\left(j\right) - x\left(j\right)\right], \\[3mm] y\left(n\right) = y\left(0\right) + \dfrac{1}{\Gamma\left(\mu\right)} \sum_{j=1}^{n-1} \dfrac{\Gamma\left(n-1-j+\mu\right)}{\Gamma\left(n-j\right)} \\[3mm] \qquad\qquad \times \left[-\beta x\left(j\right) + \left(\alpha - 1\right) y\left(j\right) - y^3\left(j\right)\right]. \end{cases} \tag{2.30}$$

To verify the effectiveness of the 0–1 test, we compare it with the results of the phase attractors and maximum Lyapunov exponents. In particular, Figures 2.8(a) and 2.8(b) illustrate the translation components in the $p - q$ plane, and the asymptotic growth rate K as a function of length n for μ, respectively. Figure 2.8(a) depicts the Brownian-like trajectories for the fractional order value $\mu = 0.95$,

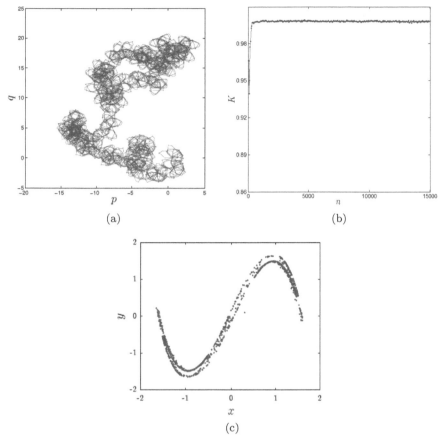

Figure 2.8. (a) $p - q$ trajectories of the 0–1 test. (b) Asymptotic growth rate K of the fractional Duffing map versus n for $\mu = 0.95$. (c) Chaotic attractor of the fractional Duffing map.

indicating that the fractional map is chaotic. On the other hand, Figure 2.8(b) shows that the asymptotic growth rate approaches 1, confirming the chaotic behavior of the map. As in Figure 2.7, the MLE is positive when $\mu = 0.95$ and $\beta = 0.2$ which confirms the chaotic behavior of the proposed map. The chaotic attractor is shown in Figure 2.8(c).

Now, we carry out the 0–1 test with fractional order $\mu = 0.8$ and system parameter $\beta = 0.5$. The translation components (p, q) are shown in Figure 2.9(a), and the asymptotic growth rate is illustrated

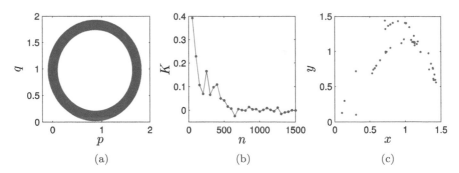

Figure 2.9. (a) Bounded $p-q$ trajectories of the 0–1 test. (b) Asymptotic growth rate K of the fractional Duffing map versus n for $\mu = 0.95$ and $\beta = 0.5$. (c) Periodic attractor of the fractional Duffing map.

in Figure 2.9(b). The bounded trajectories in Figure 2.9(a) confirm the stability of the fractional Duffing map for the fractional order $\mu = 0.95$ and system parameter $\beta = 0.5$. On the other hand, the output K approaches 0. Similarly, from the maximum LE estimation shown in Figure 2.7, it can be deduced that the fractional map can dynamically display periodic behavior when MLE approaches zero, which clearly confirms the above results.

All the results reported in this figures indicate the effectiveness of the 0–1 test in detecting chaos. Although the 0–1 test has many advantages over calculating the maximum LE, avoiding certain well-documented drawbacks of traditional tests, which will not replace traditional methods completely.

2.5. C_0 Complexity Algorithm

C_0 complexity is a new nonlinear method introduced by Shen *et al.* [56], to overcome the over-coarse graining preprocessing problem. The bigger the C_0 complexity value is, the more complex the generated sequence is, also having better anti-interference and anti-interception performance. The principle of C_0 algorithm is to divide the sequence into two components — the regular component and irregular part the proposition of irregular part is what we need, namely, C_0 complexity is calculated as the ratio of irregular component to the original signal by eliminating the regular components in the signal. Assuming a chaotic time sequence of length N as $y(n)$, $n = 1, \ldots, N-1$, the C_0 algorithm can be described as follows.

Step 1. Construct the fast Fourier transform of $y(n)$ by Y_N, as follows

$$Y_N(i) = \frac{1}{N} \sum_{n=0}^{N-1} y(n) \exp^{-2\pi \frac{nj}{N}}, \quad n = 0, 1, \ldots, N-1$$

(2.31)

where $j = \sqrt{-1}$ is an imaginary unit. Note that the power spectrum contains essential information about the nature of y.

Step 2. Define the mean square value of the amplitude spectrum $Y_N(i)$, as

$$G_N = \frac{1}{N} \sum_{j=0}^{N-1} |Y_N(i)|^2.$$

(2.32)

Next, import a control parameter r $(r > 0)$ so that C_0 complexity can better reflect the dynamic characteristics of the sequence $x(n)$. Keep all the spectrum components unchanged if their squares are less or equal to rG_N, while replace the spectrum components with zero if their square values exceed rG_N

$$\bar{Y}_N(n) = \begin{cases} Y_N(n), & \text{if } |Y_N(n)|^2 > rG_N, \\ 0, & \text{if } |Y_N(n)|^2 \leq rG_N. \end{cases}$$

(2.33)

Step 3. The inverse discrete Fourier transform of $\bar{Y}_N$ is the new time series $\bar{y}$

$$\bar{y}(n) = \sum_{j=0}^{N-1} \bar{Y}_N(i) \exp^{-2\pi \frac{nj}{N}}, \quad n = 0, 1, 2 \ldots, N-1.$$

(2.34)

$\bar{y}(n)$ reflects the regular time series of the original time series with the detailed information removed.

Step 4. The C_0 complexity is defined as the ratio area of $|y(n) - \bar{y}(n)|$ to the area of $y(n)$

$$C_0 = \frac{\sum_{n=0}^{N-1} |y(n) - \bar{y}(n)|^2}{\sum_{n=0}^{N-1} |y(n)|^2}. \qquad (2.35)$$

According to the properties demonstrated by Shen *et al.* in [56], C_0 complexity is a real number that ranges between 0 and 1 for any time series, equalling zero if the time series is constant while also approaching zero if $x(n)$ is periodic. Hence, C_0 complexity can be used as a quantitative index of complexity of a time series.

2.5.0.1. *Choice of parameters of the C_0 complexity algorithm*

The C_0 complexity depends on two parameters — r and the length of data points N. The following illustrates the selection of parameters of this algorithm by numerical experiments. To illustrate the selection of parameters of the C_0 complexity algorithm, the fractional logistic map was taken as an example [34], for which the equivalent discrete integral equation can be obtained as

$$x(x) = x(0) + \frac{1}{\Gamma(\nu)} \sum_{j=1}^{n-1} \frac{\Gamma(n-1-j+\nu)}{\Gamma(n-j)} (Kx(j)(1-x(j))x(j)).$$

$$(2.36)$$

As reported in [57–59], when $N > 2000$ the measured value is stable. Therefore, it is recommended that the sequence length is greater than 2000. Here, the parameter r is taken as $N = 7000$, the system parameter is $K = 3.5$, and the fractional order $\mu = 0.75$. The C_0 complexity is shown in Figure 2.10 where r first starts from 0.1 and gradually increases with a step value of 0.1, and then increases by 0.5 when greater than 2. It can be seen from Figure 2.10 that as the parameter r increases, the measured value of C_0 gradually increases, which shows that as r increases, the less regular parts are removed, and the measured value will increase accordingly. After that, when r is greater than 2, the measured value is stable. Hence, it is recommended that the range value of $r \in [2, 10]$, consistent with the result of the analysis in the literature [57].

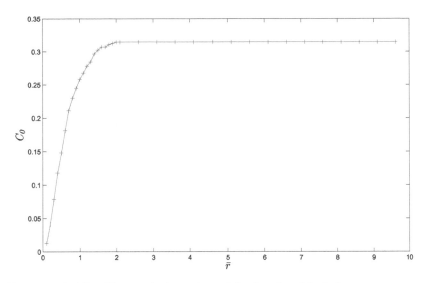

Figure 2.10. The C_0 complexity value of the fractional logistic map versus r.

2.5.1. *Applying C_0 Complexity for Analyzing the Complexity of Fractional Maps*

Now, the complexity of the fractional Duffing map is analyzed. Take $r = 10$ and $N = 5000$ here. The variation of C_0 complexity with parameter β for three fractional order values is shown in Figure 2.11. Comparing Figure 2.11 with Figure 2.7, we can see that when the Lyapunov index is less than zero, the complexity value of C_0 is zero or close to zero, indicating that the sequence generated by the system at this time is small; when the Lyapunov index is greater than zero, the C_0 complexity measure value is positively correlated with the Lyapunov index value, and the period window can be detected, indicating that the C_0 complexity can reflect the dynamics and complexity of the fractional Duffing map.

2.6. Approximate Entropy

Entropy measures the complexity of a time series by measuring the information in generating new patterns of signals. Generally, the exact entropy for a dynamical system is difficult to determine.

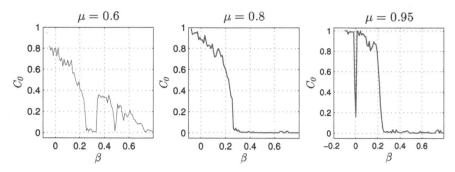

Figure 2.11. C_0 versus β of the fractional Duffing map for different fractional order values.

Therefore, Pincus [61] introduced the approximate entropy to present the complexity for both continuous and discrete dynamical systems. This method estimates the regularity where a non-negative number is assigned with higher values indicating higher complexity. The specific steps are as follows

1. Reconstruct the discrete points $x(1), x(2), \ldots, x(n)$ into one-dimensional $m - m + 1$ vectors

$$X(i) = [x(i), \ldots, x(i + m - 1)], \quad 1 \le i \le n - m + 1. \quad (2.37)$$

   These vectors are m consecutive x values which start from the first data.

2. Define the maximum absolute difference between $X(i)$ and $X(i)$, i.e.

$$d\left[X(i), X(i)\right] = \max_{0 \le k \le n-1} |x(i + k) - x(j + k)|.$$

3. Define similar tolerance r as a positive real number. Use the maximum norm to count the number K for each $i \in [1, n - m + 1]$ such that $d(X(i), X(i)) \le r$, then calculate $C_i^m(r)$ by

$$C_i^m(r) = \frac{K}{n - m + 1}. \quad (2.38)$$

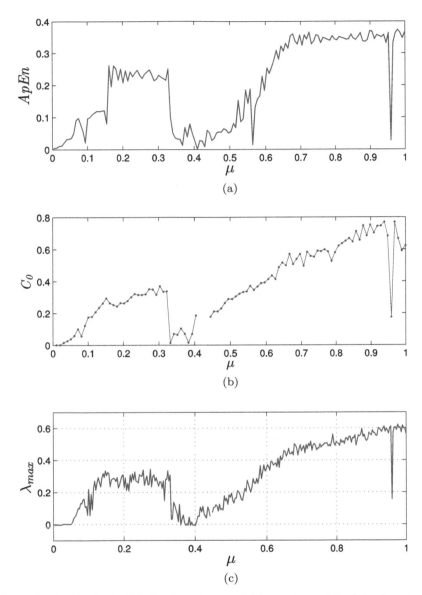

Figure 2.12. (a) *ApEn*, (b) C_0 algorithm, and (c) maximum LE of the fractional Duffing map versus fractional order μ.

4. The mean logarithm of $C_i^m(r)$, is denoted as $\phi^m(r)$, and can be calculated as

$$\phi^m(r) = \frac{1}{n-m-1} \sum_{i=1}^{n-m+1} \log C_i^m(r). \tag{2.39}$$

5. On the basis of Eq. (2.39), the approximate entropy value can be calculated as follows

$$ApEn = \phi^m(r) - \phi^{m+1}(r), \tag{2.40}$$

Pincus asserted that, for a given dynamical system, the value of the *ApEn* values depend on the two parameters, the tolerance input r and the embedding dimension m. The main purpose of embedding a time series is to unfold the projection to a state space that is representative of the original system's space, i.e. a reconstructed attractor must preserve the invariant characteristics of the original one [63]. According to Pincus' research paper [62], it is recommended to set $m = 2$ or $m = 3$, and choose the tolerance r to be 0.2 times the standard deviation (std) of the data x.

2.6.1. *Applying ApEn for Analyzing the Complexity of Fractional Chaotic Duffing Map*

Herein, the *ApEn* complexity of the fractional Duffing map (2.30) is analyzed by varying the fractional order values μ. In particular, the approximate entropy *ApEn*, the variation of C_0 complexity and maximum LEs with parameter μ are shown in Figure 2.12. It can be seen that the complexity of the fractional Duffing map (2.30) strongly depends on the variations of μ. In particular, Figure 2.12 highlights some combined values for which the approximate entropy ApEn is high, indicating that the fractional map is characterized by complex dynamical behaviors.

Chapter 3

Chaos in 2D Discrete Fractional Systems

3.1. Introduction

In the past few years, discrete dynamic behavior and its applications have attracted considerable attention in various applied fields, given the noteworthy implementations in secure communication and effective encryption. Considering the time scale theory [19], discrete fractional calculus has been useful in handling the dynamics of discrete time. This field has proved to be an effective tool in chaotic systems due to their memory effects [8]. To better understand the dynamics of discrete systems, certain implementations of the discrete fractional calculus on an arbitrary time scale are put forward and some theories are used that are connected with delta difference equations for revealing chaos modes generated by fractional order discrete maps. As a result, several chaotic maps have been established including the one-dimensional chaotic map, piecewise linear chaotic maps, and chaotic systems. These maps can be implemented during the process of image encryption. We see that the one-dimensional chaotic map has the advantage of high-level efficiency and simplicity [64]. Similarly, the two-dimensional fractional-order chaotic map can be regenerated by a more powerful process, as applied in image encryption and secure communication, by adding more parameters to its equations.

This chapter aims to present some typical examples of two-dimensional fractional order discrete-time maps with hidden chaotic dynamics of equilibrium points. In particular, some fractional order quadratic, trigonometric, and rational maps will be explored and discussed. These designed maps will be explained with their dynamical properties, by investigating the phase plots, bifurcation diagrams, largest Lyapunov exponent, and 0–1 test.

3.2. Fractional Quadratic Maps

Over the last few years, the discrete- and continuous-time chaotic dynamical systems have been widely examined, and a considerable number of research works on chaos control, chaos application, and synchronization have been deeply studied. The subject of discrete-time chaotic maps distinguished by "hidden attractors" has been only recently explored and discussed. For instance, a two-dimensional chaotic map was put forward in [65] with various kinds of stable equilibria, whereas other two-dimensional chaotic quadratic maps without equilibria and with no discontinuity in the right-hand equations were established in [66]. The authors examined methodically the complicated dynamical modes generated by these maps using several numerical simulations.

It should be noted that the aforesaid considerations are connected with the traditional chaotic maps. In more recent years, many endeavors have been dedicated to explore the fractional order quadratic maps of two types; the continuous- and discrete-time maps. Regarding the latter maps, several two-dimensional fractional order chaotic maps have been recently proposed. In this section, three types of such maps will be introduced. In particular, the two-dimensional fractional order Hénon, flow, and Lorenz maps will be examined to explore their dynamical properties.

3.2.1. *Fractional-Order Hénon Map*

The literature seems to agree that the fractional order chaotic maps possess superior properties as compared to their standard counterpart. This adds new degrees of freedom to the map's states, making them, for example, more suitable for data encryption. In particular, the variations of the fractional order value provides us a new powerful tool to characterize the dynamics of the discrete-time complex systems more deeply. In this subsection, the fractional order Hénon map will be investigated for its dynamical behaviors. In view of the Caputo fractional order difference operator, the fractional order Hénon map can be given by the following pair of the first-order difference equations:

$$\begin{cases} x_1(n+1) = 1 - \alpha x_1^2(n) + x_2(n), \\ x_2(n+1) = \beta x_1(n), \end{cases} \tag{3.1}$$

where α and β are bifurcation parameters. Consequently, we can rewrite the above two equations as follows:

$$\begin{cases} \Delta x_1(t) = 1 - \alpha x_1^2(n) + x_2(n) - x_1(n), \\ \Delta x_2(t) = \beta x_1(n) - x_2(n). \end{cases} \tag{3.2}$$

As per the discrete fractional calculus, we can modify the integer order map in its fractional order version. That is;

$$\begin{cases} \Delta_a^\nu x_1(t) = 1 - \alpha x_1^2(t - 1 + \nu) + x_2(t - 1 + \nu) - x_1(t - 1 + \nu), \\ \Delta_a^\nu x_2(t) = \beta x_1(t - 1 + \nu) - x_2(t - 1 + \nu). \end{cases}$$

$$\tag{3.3}$$

From the previous equations, we obtain the following discrete-time integral form:

$$\begin{cases} x_1(t) = x_1(0) + \dfrac{1}{\Gamma(\nu)} \sum_{s=1-\nu}^{t-\nu} (t-s-1)^{(\nu-1)} \\ \qquad \times (1 - \alpha x_1^2(s-1+\nu) + x_2(s-1+\nu) - x_1(s-1+\nu)), \\ x_2(t) = x_2(0) + \dfrac{1}{\Gamma(\nu)} \sum_{s=1-\nu}^{t-\nu} (t-s-1)^{(\nu-1)} \\ \qquad \times (\beta x_1(s-1+\nu) - x_2(s-1+\nu)). \end{cases}$$

$$(3.4)$$

As a result of this form, the numerical formula can be explicitly presented as follows:

$$\begin{cases} x_1(n) = x_1(0) + \dfrac{1}{\Gamma(\nu)} \sum_{j=1}^{n} \dfrac{\Gamma(n-j+\nu)}{\Gamma(n-j+1)} \\ \qquad \times \left(x_2(j-1) + 1 - \alpha x_1^2(j-1) - x_1(j-1) \right), \\ x_2(n) = x_2(0) + \dfrac{1}{\Gamma(\nu)} \sum_{j=1}^{n} \dfrac{\Gamma(n-j+\nu)}{\Gamma(n-j+1)} \left(\beta x_1(j-1) - x_2(j-1) \right). \end{cases}$$

$$(3.5)$$

where $x_1(0)$ and $x_2(0)$ are the initial states.

3.2.1.1. *Lyapunov exponents method*

Using Lyapunov exponents is perhaps the most important qualitative measure of chaos. From this point of view, we consider the problem of estimating these exponents for the fractional order Hénon map. System (3.5) represents two specific numerical formulas called the two-dimensional fractional-order Hénon map with two state variables x_1 and x_2. This system can display chaotic behavior for some fractional order values and can also display periodic motion for certain other values [67]. Anyway for this specific system, the

Jacobian matrix can be outlined as follows:

$$J(n) = \begin{pmatrix} a(n) & b(n) \\ c(n) & d(n) \end{pmatrix}, \tag{3.6}$$

where

$$
\begin{cases}
a(n) = a(0) + \dfrac{1}{\Gamma(\nu)} \displaystyle\sum_{j=1}^{n} \dfrac{\Gamma(n-j+\nu)}{\Gamma(n-j+1)} \\
\qquad \times \left(-2\alpha a(j-1)x(j-1) - a(j-1) + c(j-1) \right), \\[4pt]
b(n) = b(0) + \dfrac{1}{\Gamma(\nu)} \displaystyle\sum_{j=1}^{n} \dfrac{\Gamma(n-j+\nu)}{\Gamma(n-j+1)} \\
\qquad \times \left(-2\alpha b(j-1)x(j-1) - b(j-1) + d(j-1) \right), \\[4pt]
c(n) = c(0) + \dfrac{1}{\Gamma(\nu)} \displaystyle\sum_{j=1}^{n} \dfrac{\Gamma(n-j+\nu)}{\Gamma(n-j+1)} \left(\beta a(j-1) - c(j-1) \right), \\[4pt]
d(n) = d(0) + \dfrac{1}{\Gamma(\nu)} \displaystyle\sum_{j=1}^{n} \dfrac{\Gamma(n-j+\nu)}{\Gamma(n-j+1)} \left(\beta b(j-1) - d(j-1) \right),
\end{cases}
$$

$$\tag{3.7}$$

where $a(0) = b(0) = c(0) = d(0) = 1$. So, if we let $\beta = 1.3$ and $\alpha = 0.3$, then the phase portrait and the Lyapunov exponents' estimation of the fractional order Hénon map, according to some parameters and according to fractional order values, can be shown in Figure 3.1. In particular, we note that based on Figures 3.1(a) and 3.1(b), when $\nu = 0.98$ and $a = 1.02$, the system will be periodic. This is, actually, because there are few points located in the phase portraits and the largest Lyapunov exponent, converges to zero, confirming the regular dynamics for the considered system. On the other hand, if we let $a = 1.3$, $\nu = 0.96$, then the chaotic attractor and one positive Lyapunov exponent will be successfully generated as depicted in Figures 3.2(a) and 3.2(b), respectively. Besides, if we let the parameter α vary freely from 0 to 1.5 with step size of 0.0015 and with the fractional order value $\nu = 0.96$, then the largest

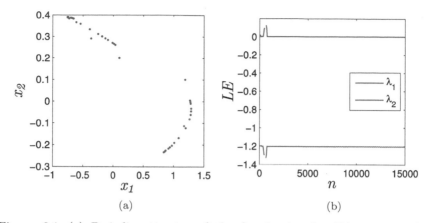

(a)

(b)

Figure 3.1. (a) Periodic attractor of the fractional order Hénon map when $\nu = 0.98$ and $a = 1.02$. (b) The corresponding Lyapunov exponents' estimation.

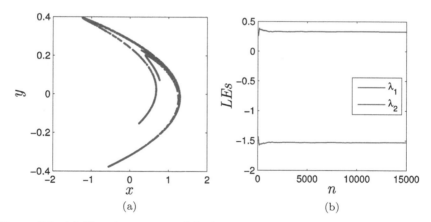

(a)

(b)

Figure 3.2. (a) Chaotic attractor of the fractional order Hénon map when $\nu = 0.96$ and $a = 1.3$. (b) The corresponding Lyapunov exponents' estimation.

Lyapunov exponent values can be generated as shown in Figure 3.3. Obviously, it can be observed that the high complexity region of the considered system is obvious when α increases. Also evident is that the proposed algorithm is really effective for detecting chaos of the fractional order discrete-time chaotic systems. Furthermore, the changes that occurred to the largest Lyapunov exponent values coincide completely with the bifurcation diagrams constructed for the same interval of changes in the control parameter α.

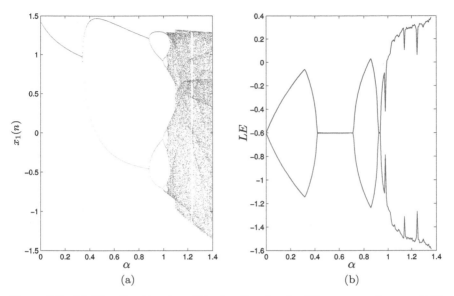

Figure 3.3. (a) Chaotic attractor of the fractional order Hénon map when $\nu = 0.96$ and $a = 1.3$. (b) The corresponding Lyapunov exponents' estimation.

3.2.1.2. *The 0–1 test method*

The sensitivity of the chaos of fractional order discrete-time maps can be clarified using the 0–1 test method. In particular, this method can be employed to identify chaos in a series of data where the phase space reconstruction is not necessary. To demonstrate the reliability of the 0–1 test method, we aim to apply it to the fractional order Hénon map for the purpose of detecting chaos generated by its dynamics.

Here, the 0–1 test can be applied directly to the solution $x(n)$ of the system's parameter $\beta = 0.3$ and for different fractional order values. However, the results are reported in Figures 3.5 and 3.4 according to $\nu = 0.98$, $\alpha = 1.02$ and $\nu = 0.96$, $\alpha = 1.3$, respectively. For instance, Figure 3.4 depicts the asymptotic growth rate K as a function of length n, and the dynamics of the translation components in the pq-plane. Since the asymptotic growth rate K approaches one and the pq-plane highlights the Brownian-like trajectories, the fractional-order Hénon map is chaotic. On the other hand, Figure 3.5 confirms that because the asymptotic growth rate K approaches zero

Fractional Discrete Chaos

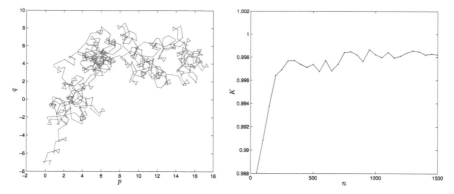

Figure 3.4. Brownian-like trajectories of translation component in the pq-plane of the fractional-order Hénon map and the corresponding translation component versus n.

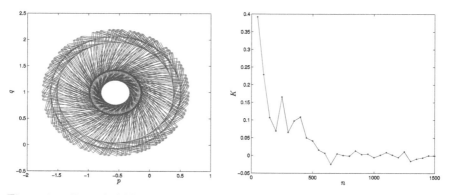

Figure 3.5. Bounded-like trajectories of translation component in the pq-plane of the fractional-order Hénon map for $\nu = 0.98$, $\alpha = 1.02$ and the corresponding translation component versus n.

in which the bounded trajectories in the pq-plane are depicted, the map at hand indicates the periodic dynamic. Besides, the 0–1 test is carried out successively by varying the parameter α from 0 to 1.40. The step size is taken here into account as 0.001 for α. The plot of the asymptotic growth rate K versus α is shown in Figure 3.6, where we observe that the fractional order Hénon map (3.5) is chaotic over most of the range $a \in (1.07, 1.40)$ when K approaches 1. At the same time, Figure 3.5 depicts the Brownian-like trajectories, confirming the chaotic dynamics of the system for $\alpha = 1.3$.

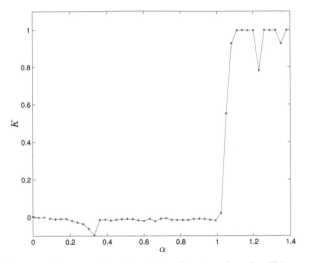

Figure 3.6. Asymptotic growth rate of the fractional order Hénon map versus α corresponding to the bifurcation diagram and Lyapunov exponents as seen in Figure 3.10.

Based on the results reported in the above discussion, one should note the effectiveness of the 0–1 test in detecting chaos overall. In fact, we find that the generated figures here are in good agreement with the Lyapunov exponents results reported in the previous subsection. Although the 0–1 test has many advantages over calculating the largest Lyapunov exponent, avoiding certain well-documented drawbacks of traditional tests, it will not replace traditional methods completely.

3.2.2. *Fractional Order Flow Map*

We now move to study the fractional order of the flow chaotic map that can be described as

$$\begin{cases} {}^{c}\Delta_{a}^{\nu}x\left(t\right) = y\left(t+\nu-1\right) + (\theta-1)x\left(t+\nu-1\right), \\ {}^{c}\Delta_{a}^{\nu}y\left(t\right) = \lambda + x^2\left(t+\nu-1\right) - y\left(t+\nu-1\right),\ t \in N_{a-\nu+1}, \end{cases} \tag{3.8}$$

where x and y stand for state variables of the fractional flow map, θ and λ are the system's parameters, and Δ_{a}^{ν} is the Caputo difference operator of the fractional order $\nu \in (0, 1]$. In fact, this discrete-time map can generally exhibit chaotic behavior. For instance,

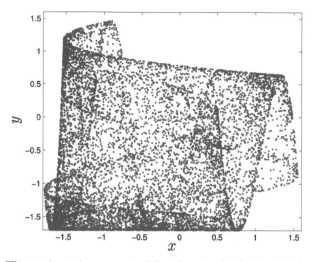

Figure 3.7. The trajectories generated by the standard discrete flow map with $\theta = -0.1$, $\lambda = -1.7$, and $(x(0), y(0)) = (0, 1)$.

when $(\theta, \lambda) = (-0.1, -1.7)$, the resultant chaotic attractor can be then depicted as in Figure 3.7. However, the numerical formula corresponding to (3.8) is given by:

$$
\begin{cases}
x(n) = x(0) + \dfrac{1}{\Gamma(\nu)} \displaystyle\sum_{j=1}^{n} \dfrac{\Gamma(\nu - j + \nu)}{\Gamma(\nu - j + 1)} (y(j-1) + (\theta - 1)x(j-1)), \\[4mm]
y(n) = y(0) + \dfrac{1}{\Gamma(\nu)} \displaystyle\sum_{j=1}^{n} \dfrac{\Gamma(\nu - j + \nu)}{\Gamma(\nu - j + 1)} (\lambda + x^2(j-1) - y(j-1)),
\end{cases}
$$

$$(3.9)$$

for $0 < \nu \le 1$.

With the same parameter values defined above and with $[x(0), y(0)] = [0, 1]$, the fractional order flow map is run for 9000 points. This, actually, generates the trajectories depicted in Figure 3.8 according to $\nu = 0.98$ and $\nu = 0.973$ in the xy-plane. Obviously, the considered map converges to a certain bounded attractor. In a similar manner to the previous fractional order maps, when the fractional order value ν decreases below the threshold 0.973, the trajectories of the map at hand will diverge to infinity. In addition, keeping $\lambda = -1.7$, $\nu = 0.98$, and $[x(0), y(0)] = [0, 1]$

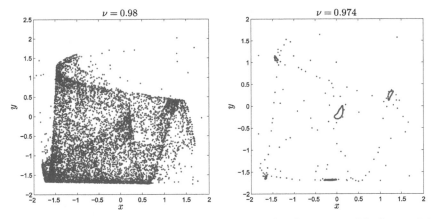

Figure 3.8. The trajectories of the fractional order flow map with $\theta = -0.1$, $\lambda = -1.7$, $(x(0), y(0)) = (0, 1)$, and with the fractional order values $\nu = 0.98$ and $\nu = 0.973$.

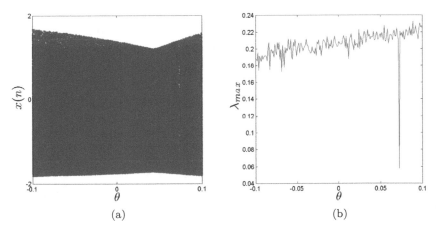

Figure 3.9. (a) Bifurcation diagram of the fractional order flow map with $\lambda = -1.7$, $(x(0), y(0)) = (0, 1)$, and with the fractional order values $\nu = 0.98$ and $\nu = 0.973$. (b) The corresponding largest Lyapunov exponent.

and continuously varying θ within the interval $[-0.1, 0.1]$ through a step size of $\Delta\theta = 0.001$ yield the bifurcation diagram depicted in Figure 3.9. This clearly shows that the fractional order flow map is indeed chaotic in the interval $[-0.1, 0.1]$, which agrees completely with the largest Lyapunov exponent as seen in Figure 3.9(b).

3.2.3. *Fractional Order Lorenz Map*

The discretization of the well-known continuous-time Lorenz chaotic system was carried out in [68]. This yielded the integer order discrete-time Lorenz chaotic map that has the following form:

$$\begin{cases} x\,(n+1) = (1+\gamma\delta)\,x\,(n) - \delta y\,(n)\,x\,(n), \\ y\,(n+1) = (1-\delta)\,y\,(n) + \delta x^2\,(n), \end{cases} \tag{3.10}$$

where γ and δ are some parameters that can typically determine the general behavior and the dynamics of the system. For instance, this system has a chaotic attractor when $\gamma = 1.25$ and $\delta = 0.75$. By tracking the same steps as mentioned in the previous section, the fractional-order Lorenz map can be formulated as follows:

$$\begin{cases} {}^{C}\Delta_a^{\nu}x\,(t) = \gamma\delta x\,(t-1+\nu) - \delta y\,(t-1+\nu)\,x\,(t-1+\nu), \\ {}^{C}\Delta_a^{\nu}y\,(t) = \delta\left(-y\,(t-1+\nu) + x^2\,(t-1+\nu)\right), \end{cases} \tag{3.11}$$

for $t \in N_{a-\nu+1}$ and $0 < \nu \leq 1$. In fact, the above map is symmetrical about x-axis since it remains unchanged when the transformation $(-x, y) \to (x, y)$ is applied. However, the numerical formula of the considered map can be consequently given as

$$\begin{cases} x\,(n) = x\,(a) + \dfrac{1}{\Gamma\,(\nu)} \displaystyle\sum_{j=1}^{n} \dfrac{\Gamma\,(n-j+\nu)}{\Gamma\,(n-j+1)} \\ \qquad\quad \times\,(\gamma\delta x\,(j-1) - \delta y\,(j-1)\,x\,(j-1)), \\ y\,(n) = y\,(a) + \dfrac{1}{\Gamma\,(\nu)} \displaystyle\sum_{j=1}^{n} \dfrac{\Gamma\,(n-j+\nu)}{\Gamma\,(n-j+1)}\,(\delta(-y\,(j-1) + x^2\,(j-1)). \end{cases}$$

$$\tag{3.12}$$

In the following text, we will consider the system's parameters as $(\gamma, \delta) = (1.25, 0.75)$, the fractional order value as $\nu = 1$, $a = 0$, and also the initial conditions as $(x(0), y(0)) = (0.1, 0)$. As a result, one notices that the solution of the fractional order Lorenz map (3.11) can approach the strange attractor as depicted in Figure 3.10(b). At the same time, Figure 3.11 shows the trajectories of such maps obtained with 9000 points and with taking the

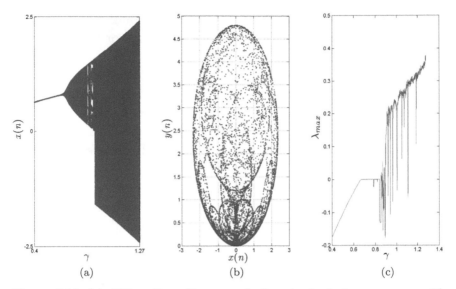

Figure 3.10. (a) Bifurcation diagram of the standard Lorenz map with $(x(0), y(0)) = (0.1, 0)$ and $(\gamma, \delta) = (1.25, 0.75)$. (b) The solutions plotted in xy-plane.

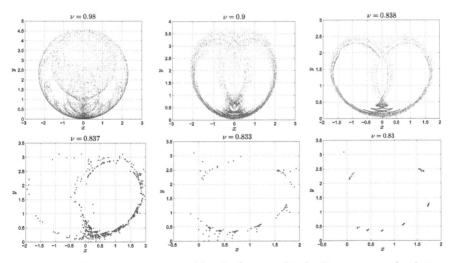

Figure 3.11. Trajectories generated by the fractional order Lorenz map after being run for 9000 points with $(x(0), y(0)) = (0.1, 0)$, $(\gamma, \delta) = (1.25, 0.75)$, and different fractional order values ν.

same parameters' values and the same initial conditions that were previously mentioned, but with different fractional order values here given as $\nu \in \{0.98, 0.838, 0.837, 0.833, 0.83\}$. Note that all solutions, in all six cases, are plotted in the xy-plane. However, it can be inferred based on the previous discussion that a slight change in the fractional order value ν can lead to transition from a strange attractor to two invariant circles and then to seven periodic orbits. These changes denote that the strange attractor will be destroyed as ν decreases.

In order to study the effect of the fractional order value on the dynamics of the chaotic system, we should consider its bifurcation analysis. For accurate and stable outputs, we calculate the bifurcation for 2000 points. The initial states are again fixed at $(x(0), y(0)) = (0.1, 0)$, and the parameter δ at $\delta = 0.75$. To see the gained numerical results, Figure 3.12 shows the bifurcation graphs obtained by taking the parameter γ as the critical parameter, according to different fractional order values. We note that as ν decreases, the bifurcation diagram changes. Particularly, for $\nu = 0.98$, the fractional order map converges to a fixed point over the approximate interval $\gamma \in [0.4, 0.6429]$. For $\gamma \in [0.92, 1.27]$, the trajectories of the considered map converge to a chaotic attractor. However, for $\nu = 0.9$, the fixed point convergence and the chaotic attractor intervals expand to $\gamma \in [0.4, 0.7726]$ and $\gamma \in [1.1, 1.27]$, respectively. Finally, for $\nu = 0.83$, we see that a fixed point convergence expands to γ between 0.4 and

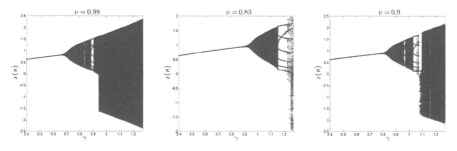

Figure 3.12. Bifurcation diagrams of the fractional order Lorenz map with $(x(0), y(0)) = (0.1, 0)$ and $\delta = 0.75$ according to the fractional order values $\nu = 0.98$, $\nu = 0.9$, and $\upsilon = 0.83$.

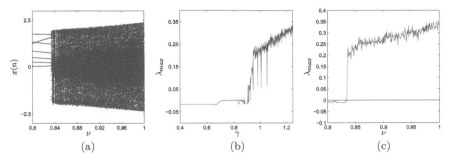

Figure 3.13. (a) Bifurcation diagram of the fractional order Lorenz map with $(x(0), y(0)) = (0.1, 0)$, $(\gamma, \delta) = (1.25, 0.75)$, and considering ν as the critical parameter; (b) The largest Lyaponov exponent, considering the fractional order value ν as the bifurcation parameter; (c) The largest Lyapunov exponent, considering γ as the bifurcation parameter and $\nu = 0.98$.

0.8886, whereas a small chaotic interval expands to $\gamma \in [1.255, 1.27]$ approximately. Comparing these results to the bifurcation diagram obtained in accordance with $\nu = 1$, as in Figure 3.11(a), leads to the conclusion that the bifurcation diagrams of the fractional order map are stretched along the γ-axis.

From the previous discussion, it can be inferred that the fractional order quadratic maps show rich dynamical behaviors. Thus, it may be declared here that such maps can undoubtedly have the potential to be useful in many applied science and engineering fields. This assertion comes certainly from the fact that those maps have more complexity than that of integer order ones, especially when the fractional order values are small. Just to name a few, due to the rich complex dynamical behaviors, the proposed fractional order maps are deemed more suitable for secure communication and image encryption than other integer order maps.

3.3. Fractional Trigonometric Maps

As of now, the discrete chaotic map can generate rich and complex dynamics. It is not only sensitive to minor disruptions to initial conditions and parameters' values, but also to the change in the fractional order values. As a result, the simple forms of fractional order discrete maps that can exhibit rich dynamics are more appropriate for secure communication and data encryption. From this perspective, the study of certain fractional order discrete maps including the fractional order trigonometric maps is essential and significant for developing the dynamics of fractional calculus. This section puts forward two types of fractional order trigonometric maps; the fractional order sine map and the fractional order sine–sine map. In particular, these two maps will be examined by their dynamical properties. This will be performed through investigating the phase plots, bifurcation diagrams, largest Lyapunov exponent, and 0–1 test.

3.3.1. *Fractional Order Sine Map*

This part will follow the same procedure as mentioned in the earlier subsections. By replacing the term x^2 from the Hénon map with the trigonometry $\sin(x)$ term, a new two-dimensional iterated map will be, as proposed in [69], established. This map has the form:

$$\begin{cases} x\,(n+1) = 1 - \alpha \sin x\,(n) + \beta y\,(n), \\ y\,(n+1) = x\,(n), \end{cases} \tag{3.13}$$

where x, y are the states of the discrete-time system and α, β are some bifurcation parameters. In fact, this map can yield rich and complex dynamics, as yielding a C^∞ mapping, considered as a generalization of the chaotic attractor. To be more clear, this generalization is done with the so-called "multi-folds" by a period-doubling bifurcation route to chaos. According to [69], the considered map exhibits chaotic behavior when α falls in the interval $[-150, 200]$ and $\beta = 0.3$. Namely, for $\alpha = 3.8$, the phase space is plotted in Figure 3.14, confirming the existence of chaos.

In what follows, we aim to study the dynamics of the fractional order map in the sense of the Caputo difference operator that corresponds to the integer order sine map (3.13). To this aim, we will present some numerical simulations for a sufficient understanding of the behavior types generated from the effect of the fractional order value on the dynamics of the integer order map. In a similar manner to the previous subsections, to derive the fractional order map in

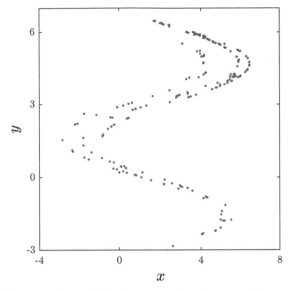

Figure 3.14. Phase portrait of the integer order sine map for parameter values $(\alpha, \beta) = (3.8, 0.3)$.

view of the Caputo difference operator from the integer order map (3.13), we start by applying the delta difference operator to the latter map, as follows:

$$\begin{cases} \Delta x\,(n) = 1 - \alpha \sin x\,(n) + \beta y\,(n) - x\,(n), \\ \Delta y\,(n) = x\,(n) - y\,(n). \end{cases} \quad (3.14)$$

Consequently, applying the Caputo delta difference operator to the above system yields the fractional-order sine map, which can be given as

$$\begin{cases} ^{C}\Delta_a^\nu x\,(t) = 1 - \alpha \sin x\,(t + \nu - 1) + \beta y\,(t + \nu - 1) - x\,(t + \nu - 1), \\ ^{C}\Delta_a^\nu y\,(t) = x\,(t + \nu - 1) - y\,(t + \nu - 1), \end{cases}$$
$$(3.15)$$

where $t \in \mathbb{N}_{a+1-\nu}$, $0 < \nu < 1$, and a is the starting point.

In order to investigate the dynamics of the fractional order sine map, we consider here the equivalent discrete formula for system

(3.15) as:

$$
\begin{cases}
x\left(n\right) = x\left(0\right) + \dfrac{1}{\Gamma\left(\nu\right)} \displaystyle\sum_{j=1}^{n} \dfrac{\Gamma\left(n-j+\nu\right)}{\Gamma\left(n-j+1\right)} \\
\qquad \times \left(1 - \alpha\sin x\left(j-1\right) + \beta y\left(j-1\right) - x\left(j-1\right)\right), \\[2mm]
y\left(n\right) = y\left(0\right) + \dfrac{1}{\Gamma\left(\nu\right)} \displaystyle\sum_{j=1}^{n} \dfrac{\Gamma\left(n-j+\nu\right)}{\Gamma\left(n-j+1\right)} \left(x\left(j-1\right) - y\left(j-1\right)\right),
\end{cases}
$$

$$(3.16)$$

where $0 < \nu \leq 1$. As per the above discrete-time formula, the proposed fractional order map (3.15) highlights the feature of the memory effects. Such prevalent features, yielded from applying fractional calculus, can indicate that the iterated solutions x_n and y_n can be determined by all previous states. However, in the following segment, we intend to study the effect of the fractional order value ν on the dynamics of the fractional order sine map (3.15) through implementing formula (3.16). For several upcoming numerical simulations, we will consider the same bifurcation parameter values as used previously.

3.3.1.1. *Bifurcation diagrams, Lyapunov exponents and phase portraits*

First of all, the bifurcation diagram can be depicted in Figure 3.15 whereby $\nu = 1$, and α is kept within the interval $[-1, 4]$ and varied in the steps of $\Delta\alpha = 0.003$. Since $\nu = 1$, the proposed fractional order map is expected to be reduced to its corresponding standard one with its solution $x\left(n\right)$ depending on all past information $x\left(n-1\right), x\left(n-2\right), \ldots, x\left(0\right)$. Once the fractional order value ν is varied from 1 to some smaller values, it then has an impact on the dynamics of the map. In such way, the phase portraits for $\nu = 0.976$, $\nu = 0.78$, and $\nu = 0.65$ are shown in Figure 3.16. The corresponding bifurcation diagrams are also depicted in Figure 3.17 for $\alpha \in [-1, 4]$. It is observed that if we fix the two bifurcation parameters α and β, and frequently change the fractional-order value ν, the bounded attractor of the fractional-order map will be distributed in several

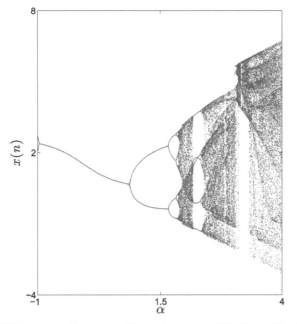

Figure 3.15. Bifurcation diagram of the sine map with the critical parameters $-1 \leq \alpha \leq 4$ and $\nu = 1$.

large regions. In the meantime, the transient state can be observed in Figure 3.18 for $\nu = 0.63$. This, actually, emphasizes that the solution of the considered system converges to a bounded attractor at the beginning, and then diverges gradually to infinity in a different direction.

Going back to the bifurcation diagram of the standard sine map shown in Figure 3.15, we observe that such plot agrees completely with what was reported in [69]. In addition, we note that the standard map goes from a fixed point to a series of period-doubling bifurcations throughout the interval $(0.76, 1.86]$. Also, it then exhibits a chaotic attractor for $\alpha \in (1.86, 2.16]$. At the same time, we note that once ν passes to the interval $(2.16, 2.27]$, the map goes back to a fixed point for $\alpha \in (2.27, 2.39]$. There are also several periodic windows for $\alpha >$ 2.92. In Figure 3.17, we observe that as ν is made smaller, the fixed point and periodic orbit phases disappear in the intervals $(2.16, 2.27]$ and $(2.27, 2.39]$. Besides, an increase in the chaotic attractors is also

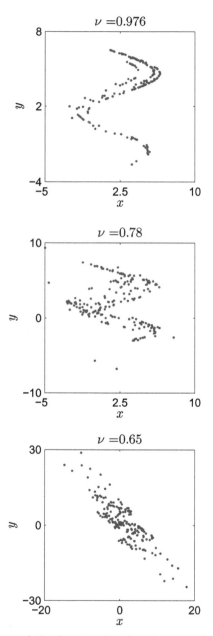

Figure 3.16. Attractors of the fractional order sine map for different fractional order values ν.

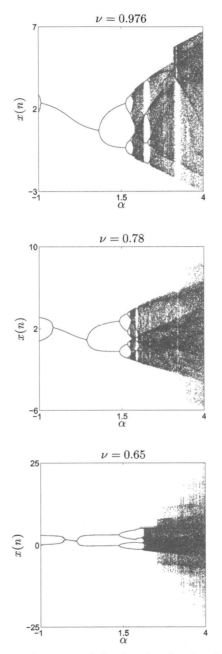

Figure 3.17. Bifurcation diagrams of the fractional order sine map for different fractional order values ν.

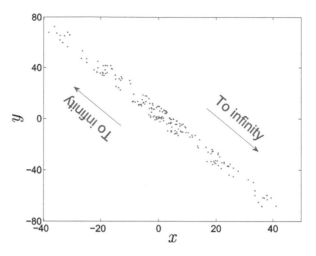

Figure 3.18. The transient state of the fractional order sine map for $\nu = 0.63$.

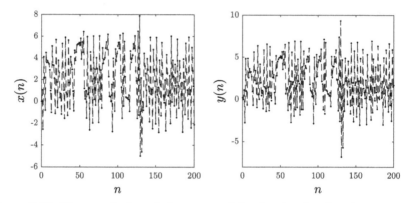

Figure 3.19. Time evolution of the states of the fractional order sine map with $\nu = 0.78$.

observed. For completeness, Figure 3.19 shows the time evolution of the chaotic states for $\nu = 0.78$.

To further visualize the influence of the fractional order value ν on the dynamic behavior of the new map (3.15), the dynamics of the fractional order sine map is furthermore analyzed by continuously varying the fractional order value ν. Since the Lyapunov exponent appears to be a good indicator for exploring the existence of chaos, we compute the Lyapunov characteristic exponents via the Jacobian

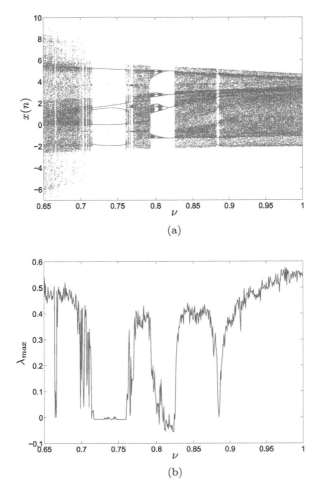

(a)

(b)

Figure 3.20. Bifurcation: (a) The largest Lyapunov exponent (b) Diagram for the fractional order sine map with respect to ν and with $\alpha = 3$, $\beta = 0.3$.

method. To this aim, the corresponding bifurcation diagram and the largest Lyapunov exponent diagrams taken over the range $\nu \in [0.65, 1]$ are shown in Figure 3.20 according to the parameter set $(\alpha, \beta) = (3, 0.3)$. As the value of ν decreases below 1, the fractional order sine map (3.15) can then be observed to undergo a transition from chaotic to periodic states, and then it gradually becomes chaotic again at 0.7124. Furthermore, as ν continues to decrease, the system eventually converges to an unbounded attractor. However, all of the above results can lead to the conclusion that the fractional order value ν can be taken as a bifurcation parameter.

3.3.1.2. *The 0–1 test method*

In order to prove the existence of chaotic behavior of the fractional order sine map, the 0–1 test method is considered along with the system's parameters $(\alpha, \beta) = (3.8, 0.3)$. The 0–1 test is a binary test constructing from a random walk type process with time series data. A simple visual test is done by plotting the trajectories of the translation components in the pq-plane. Generally, the unbounded trajectories in the pq-plane typically imply chaotic behavior, whereas the bounded ones imply a regular behavior. Here, we have applied the test directly to the solution $x(n)$. The results are, however, depicted in Figures 3.21 and 3.22. In particular, Figure 3.21 illustrates the dynamics of the translation components (p, q). Clearly, the unbounded trajectories in the pq-plane correspond to the chaotic state. At the same time, Figure 3.22 shows that the asymptotic growth rate K approaches 1 as n increases, indicating chaotic behavior of the system on hand.

3.3.2. *Fractional Order Sine–Sine Map*

We now move to a new two-dimensional chaotic map that includes two sine terms in construction. The so-called sine-sine map was proposed and studied in [70], and defined by:

$$\begin{cases} x\,(n+1) = \sin x\,(n) - \sin 2y\,(n), \\ y\,(n+1) = x\,(n), \end{cases} \tag{3.17}$$

where x and y are two dependent state variables. Although this map is simpler than the previous one, it exhibits richer dynamics with a class of basin, excluding a set of measure zero. The phase space of this map can be observed in Figure 3.23 according to the initial conditions $[x\,(0), y\,(0)] = [1, 1]$. Based on this figure, we observe that the shape of the attractor agrees completely with the results reported in [70].

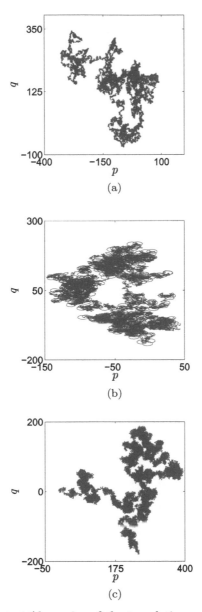

Figure 3.21. The 0–1 test (dynamics of the translation components p and q) of the fractional order sine map for different fractional order values (a) $\nu = 0.976$, (b) $\nu = 0.78$, (c) $\nu = 0.65$.

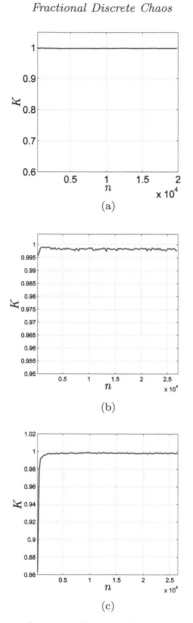

Figure 3.22. The 0–1 test (asymptotic growth rate versus n) of the fractional order sine map for different fractional order values (a) $\nu = 0.976$, (b) $\nu = 0.78$, (c) $\nu = 0.65$.

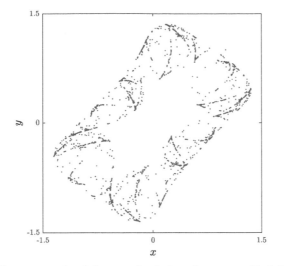

Figure 3.23. Phase portrait of the standard sine-sine map with initial conditions $[x(0), y(0)] = [1, 1]$.

Similar to the previous subsections, we use the discrete fractional calculus notations for considered map in its fractional order form. This, however, would give the following form:

$$\begin{cases} {}^C\Delta_a^\nu x(t) = \sin x\,(t+\nu-1) - \sin 2y\,(t+\nu-1) - x\,(t+\nu-1), \\ {}^C\Delta_a^\nu y(t) = x\,(t+\nu-1) - y\,(t+\nu-1), \end{cases}$$

$$(3.18)$$

where $t \in \mathbb{N}_{a+1-\nu}$ and $0 < \nu < 1$. Accordingly, the equivalent numerical formula of the above system can then be indicated as

$$\begin{cases} x(n) = x(0) + \dfrac{1}{\Gamma(\nu)} \displaystyle\sum_{j=1}^{n} \dfrac{\Gamma(n-j+\nu)}{\Gamma(n-j+1)} \\ \qquad\qquad \times (\sin x\,(j-1) - \sin 2y\,(j-1) - x\,(j-1)), \\ y(n) = y(0) + \dfrac{1}{\Gamma(\nu)} \displaystyle\sum_{j=1}^{n} \dfrac{\Gamma(n-j+\nu)}{\Gamma(n-j+1)} (x\,(j-1) - y\,(j-1)), \end{cases}$$

$$(3.19)$$

where $x(0)$ and $y(0)$ are the initial conditions. In order to proceed with this subject, we will present in the following segments several numerical results that coincide with other similar results reported in [71].

3.3.2.1. *Dynamics of the fractional order sine–sine map*

Using formula (3.19) together with two different fractional order values $\nu = 0.989$ and $\nu = 0.976$ yield the phase space portraits as seen in Figure 3.24. The largest Lyapunov exponent for these two

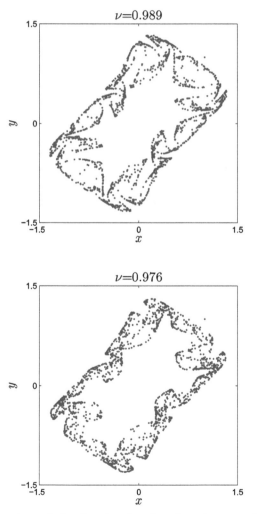

Figure 3.24. Attractors of the fractional order sine–sine map for two different fractional order values $\nu = 0.989$ and $\nu = 0.976$.

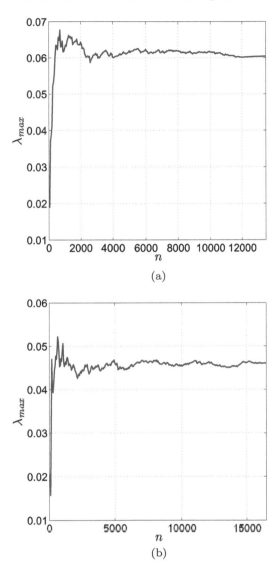

Figure 3.25. The largest Lyapunov exponent of the fractional order sine–sine map for different fractional order values: (a) $\nu = 0.989$, and (b) $\nu = 0.976$.

orbits are also illustrated in Figure 3.25. Since the fractional order map has a positive largest Lyapunov exponent, the phase portrait is therefore a chaotic attractor as seen in Figure 3.24. The calculated values of K of the 0–1 test for these two orbits are also presented in

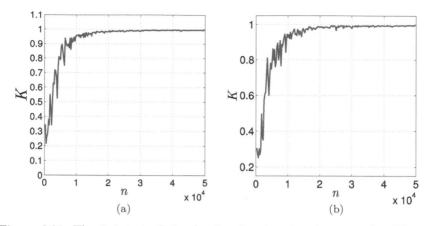

Figure 3.26. The 0–1 test of the fractional order sine–sine map for different fractional order values: (a) $\nu = 0.989$ and (b) $\nu = 0.976$.

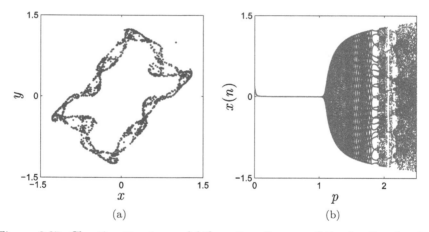

Figure 3.27. Chaotic attractor and bifurcation diagram of the fractional order sine–sine map for $\nu = 0.976$ and for the critical parameter p varying from 0 to 2.5: (a) chaotic attractor, (b) bifurcation diagram.

Figure 3.26. We observe that K approaches 1 when $\nu = 0.989$ and $\nu = 0.976$. Therefore, the 0–1 test confirms the existence of chaos. Setting $\nu = 0.976$ and continuously varying the parameter p in the interval $[0, 2.5]$ by steps of $\Delta p = 0.001$ yield the bifurcation diagram as in Figure 3.27. To further analyze the effect of the fractional order

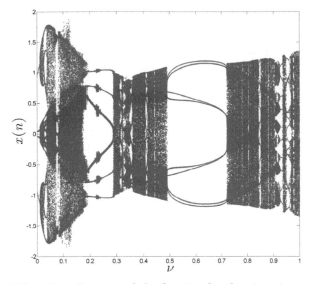

Figure 3.28. Bifurcation diagram of the fractional order sine–sine map with the critical parameter ν for two different initial conditions: (red) $(x_0, y_0) = (-1, -1)$ and (blue) $(x_0, y_0) = (1, 1)$.

value ν on the dynamic behavior of the fractional order sine–sine map (3.17), we continuously vary ν from 0 to 1 in steps of 0.001, and observe the behavior of the map. The bifurcation diagrams according to two different initial conditions can be then seen in Figure 3.28, where the red diagram corresponds to the initial conditions $(-1, -1)$, while the blue diagram corresponds to the initial conditions $(1, 1)$. It is clear that the fractional order map (3.17) is invariant under the transformation from (x, y) to $(-x, -y)$. Based on Figure 3.28, we find that even with a fractional order value, the map still exhibits chaotic behavior. The chaotic properties of this map disappear in the interval $\nu \in [0.952, 0.967]$, and then reappear for $\nu \in [0.939, 0.951]$ as seen in Figure 3.28. Finally, we note that when $\nu \leq 0.03$, chaos disappears once more and then the considered system becomes stable.

3.4. Fractional Rational Maps

The category of rational chaotic maps includes those for which the nonlinearity is a rational function of the states of the system.

The rational maps have recently become a hot topic since they confirm that such maps can generate a complex chaotic mode as revealed in several studies. Interest in this kind of maps began with one-dimensional maps as reported in [72], where the dynamics of a one-dimensional rational map were showed to generate more complicated behaviors than that of the standard logistic map. However, additional investigation in this field yielded several two-dimensional maps comprising the Rulkov map [73], Chang *et al.* map [74]. The latter two maps with fractional order will be examined here with respect to their dynamics and the impact of the fractional order value on the range and type chaotic behaviors.

3.4.1. *Fractional Order Rulkov Map*

About a few decades ago, a new discrete iterative system with rational fraction was discovered in the study of evolutionary algorithms. This map is the Rulkov map which is a two-dimensional discrete-time model with rational fraction and has the form [75]:

$$\begin{cases} x\,(n+1) = \dfrac{\mu}{1 + x^2\,(n)} + y\,(n), \\ y\,(n+1) = y\,(n) - \sigma x\,(n) - \theta, \end{cases} \qquad (3.20)$$

where $x\,(n)$ and $y\,(n)$ denote the states of the map, respectively representing the fast and slow dynamical variables, whereas θ and σ represent certain positive parameters. To proceed with some numerical investigations, we fix the parameters σ and θ as well as vary the value of μ in the interval $[0, 5]$ with step-size $\Delta\mu = 0.007$. Accordingly, the bifurcation diagram can be then generated and plotted as shown in Figure 3.29.

For modifying the system (3.20), a new fractional order version can be proposed as given in [76] as:

$$\begin{cases} {}^C\Delta_a^\nu x\,(t) = \dfrac{\mu}{1 + x^2\,(t - 1 + \nu)} + y\,(t - 1 + \nu) - x\,(t - 1 + \nu), \\ {}^C\Delta_a^\nu y\,(t) = -\sigma x\,(t - 1 + \nu) - \theta, \end{cases}$$

$$\qquad (3.21)$$

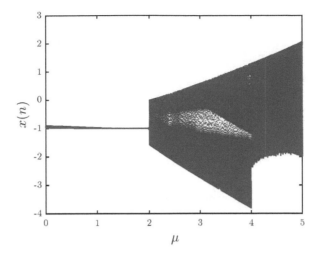

Figure 3.29. Bifurcation diagram of the integer order Rulkov rational map with $\sigma = \theta = 0.001$ and $\mu \in [0, 5]$.

for $t \in N_{a+1-\nu}$. The numerical formula, which corresponds to system (3.21), can be then expressed as follows:

$$
\begin{cases}
x(n) = x(0) + \dfrac{1}{\Gamma(\nu)} \displaystyle\sum_{j=1}^{n} \dfrac{\Gamma(n-j+\nu)}{\Gamma(n-j+1)} \\
\qquad \times \left(\dfrac{\mu}{1+x^2(j-1)} + y(j-1) - x(j-1) \right), \\
y(n) = y(0) + \dfrac{1}{\Gamma(\nu)} \displaystyle\sum_{j=1}^{n} \dfrac{\Gamma(n-j+\nu)}{\Gamma(n-j+1)} (-\sigma x(j-1) - \theta).
\end{cases} \tag{3.22}
$$

In the following numerical content, the parameter values of the above formulas are considered as: $\mu = 4.3$ and $\sigma = \theta = 0.001$. This, immediately, generates the phase portrait depicted in Figure 3.30 for $\nu = 1$ and $[x(0), y(0)] = [0.1, 0.2]$. In this case, the map is identical to the classical Rulkov model (3.20). On the other hand, Figure 3.31 shows the bifurcation diagram by taking μ as a critical parameter and $\nu = 0.85$. As per the figure, we observe that chaos exists in this map, and that the factional order value has a clear effect on the system's dynamics.

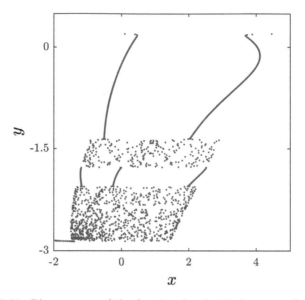

Figure 3.30. Phase space of the fractional order Rulkov map for $\nu = 1$.

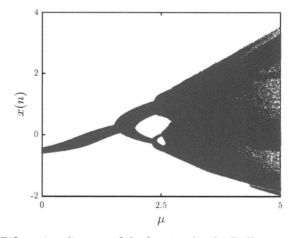

Figure 3.31. Bifurcation diagram of the fractional order Rulkov map for $\nu = 0.85$.

3.4.2. *Fractional Order Chang et al. Map*

In this part, we consider another rational discrete map that was proposed in [77]. In general, the Chang *et al.* map is given as

$$
\begin{cases}
x(n+1) = \dfrac{1}{x^2(n) + 0.1} - py(n), \\[2mm]
y(n+1) = \dfrac{1}{y^2(n) + 0.1} + qx(n),
\end{cases}
\tag{3.23}
$$

where $p, q \in [0, 0.99]$ and $x(n), y(n) \in [-1000, 1000]$. It has been shown, by using certain numerical analysis, that this map is chaotic for $p = 0.99$ and $q = 0.9$, and the bifurcation diagram and phase portraits are shown in Figure 3.32.

By using the Caputo fractional order difference operator, we can formulate the fractional order Chang *et al.* map as follows:

$$
\begin{cases}
{}^{C}\Delta_a^{\nu} x(t) = \dfrac{1}{x^2(t - 1 + \nu) + 0.1} - py(t - 1 + \nu) - x(t - 1 + \nu), \\[2mm]
{}^{C}\Delta_a^{\nu} y(t) = \dfrac{1}{y^2(t - 1 + \nu) + 0.1} + qx(t - 1 + \nu) - y(t - 1 + \nu),
\end{cases}
\tag{3.24}
$$

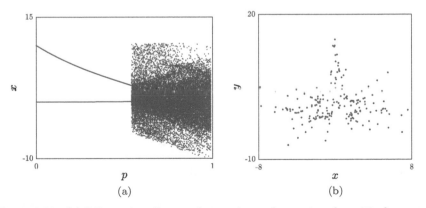

Figure 3.32. (a) Bifurcation diagram by continuously varying the critical parameter value p from 0 to 0.99. (b) Phase portrait for the Chang *et al.* map with parameter values $p = 0.99$ and $q = 0.9$.

where $t \in N_{a+1-\nu}$ and $0 < \nu \leq 1$. For $a = 0$, the discrete solution of the fractional order map (3.24) can then be obtained as:

$$\begin{cases} x(n) = x(0) + \dfrac{1}{\Gamma(\nu)} \displaystyle\sum_{j=1}^{n} \dfrac{\Gamma(n-j+\nu)}{\Gamma(n-j+1)} \\ \qquad \times \left(\dfrac{1}{x^2(j-1)+0.1} - py(j-1) - x(j-1) \right), \\ y(n) = y(0) + \dfrac{1}{\Gamma(\nu)} \displaystyle\sum_{j=1}^{n} \dfrac{\Gamma(n-j+v)}{\Gamma(n-j+1)} \\ \qquad \times \left(\dfrac{1}{y^2(j-1)+0.1} + qx(j-1) - y(j-1) \right). \end{cases} \qquad (3.25)$$

In what follows, we choose to fix $p = 0.99$ and $q = 0.9$ and set the initial conditions as $[x(0), y(0)] = [1, 1]$. By implementing numerical formula (3.25) through a certain MATLAB script, we can examine the general behavior of the fractional order Chang *et al.* map (3.24) in terms of its dependence on the fractional order value ν. To this aim and to get a general idea about the nonlinearity of its solution, the phase space is shown in Figure 3.33, where it can be inferred that the attractor changes its shape as the fractional order value ν is varied. Such changes confirm that the fractional order value has a clear effect on the dynamics of the solution. However, to further investigate the chaotic behavior, the bifurcation diagram is considered when the parameter p is varied based on the step size $\Delta p = 0.001$ over the interval $[0, 0.99]$. In particular, the bifurcation diagrams are plotted for $\nu = 0.95, 0.85, 0.7, 0.64$ as shown in Figure 3.34. This instantaneously confirms the existence of chaos. In addition, as the fractional-order value drops below 0.613, chaos disappears completely.

3.5. Fractional Unified Maps

Chaotic maps are considered to be the better solution while investigating the security of the shared information with a secret key under progressive visual cryptography considered as random in nature [78]. The fractional order unified maps are proposed for effective encryption in commercial, strategic, and personal computer-based

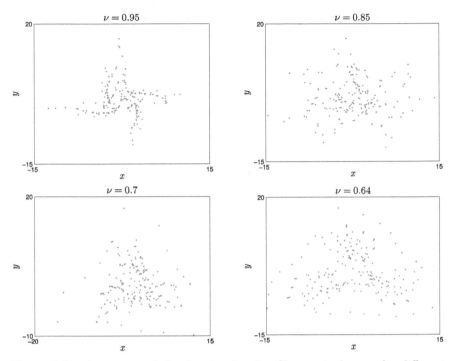

Figure 3.33. Attractors of the fractional-order Chang *et al.* map for different fractional order values.

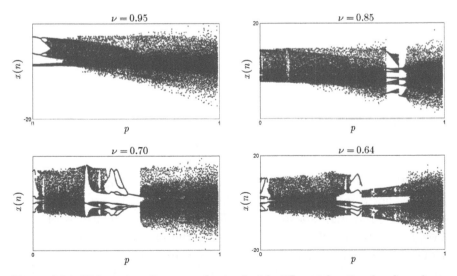

Figure 3.34. Bifurcation diagrams obtained with different fractional order values.

communications. The simulation experiments for such maps can be conducted using the evaluation metrics of entropy and mean square scheme. These maps can facilitate the process of encoding the secret image through chaotic modes generated by their dynamics [79]. They are responsible for partitioning the secrets into blocks such that the benefit of imperceptibility is included in the encoding process. In addition, such maps can also improve the strength of the encoding process such that leakage of information is avoided to the maximum [80]. In this part, we intend to handle two types of the fractional unified maps — the fractional order Hénon–Lozi type map and the fractional order Zeraoulia–Sprott map.

3.5.1. *Fractional Order Hénon–Lozi Type Map*

Motivated by the Lozi work, Alaoui *et al.* [81] proposed and analyzed a new 2D chaotic piecewise map with a form very similar to the Hénon map, but with C^1-smooth function. This map is not only better in the mathematical analysis, but it can also process the characteristic of the Hénon dynamics. Therefore, it is meaningful to extend the Hénon–Lozi type map into its fractional order case and investigate the resultant system's dynamics. To this aim, let us first consider the integer order Hénon–Lozi map which can be given as:

$$\begin{cases} x\,(n+1) = 1 - \alpha S_\varepsilon\,(x\,(n)) + y\,(n), \\ y\,(n+1) = \beta x\,(n), \end{cases} \tag{3.26}$$

where $\alpha, \beta \in \mathbb{R}$ such that $\beta \neq 0$, and S_ε is defined by:

$$S_\varepsilon : \mathbb{R} \to \mathbb{R}, \quad x \to S_\varepsilon(x) = \begin{cases} |x| & \text{if } |x| \geq \varepsilon, \\ \left(\dfrac{x^2}{2\varepsilon}\right) + \left(\dfrac{\varepsilon}{2}\right) & \text{if } |x| \leq \varepsilon, \end{cases} \tag{3.27}$$

where $0 < \varepsilon < 1$. The only difference between Lozi (or Hénon's) map is that the term $|x|$ (or x^2) is replaced by $S_\varepsilon(x)$. It was found that the considered map shares many of the qualitative features of both Lozi's and Hénon maps. For instance, the chaotic attractor of the Hénon–Lozi map under the system's parameters $(\alpha, \beta) = (1.7, 0.5)$ and with the initial conditions $(x\,(0), y\,(0)) = (1, 0)$ can be shown

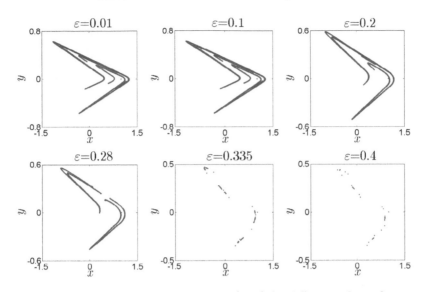

Figure 3.35. Phase plots of the map (3.26) for different values of ε.

in Figure 3.35. As with the Lozi map, zooming into a particular part of the attractors reveals clearly the fractal structure of the map. In order to gain a comprehensive understanding of the dynamics of a chaotic system, it is always helpful to examine the bifurcation diagram corresponding to a specific critical parameter. For this purpose, we have produced the bifurcation diagram of system (3.26) with $\alpha \in [0, 1.8]$, $\beta = 0.5$ and $\varepsilon = 0.1$, as shown in Figure 3.36. As it can be seen, the Hénon–Lozi map enters into chaos via period doubling bifurcation.

In the following, we aim to consider a new fractional order version of the Hénon–Lozi map formulated as per the Caputo difference operator. Such fractional order version of system (3.26) can be defined as

$$\begin{cases} {}^{C}\Delta_a^\nu x(t) = 1 - \alpha S_\varepsilon \left(x \left(t - 1 + \nu \right) \right) + y(t - 1 + \nu) - x(t - 1 + \nu), \\ {}^{C}\Delta_a^\nu y(t) = \beta x(t - 1 + \nu) - y(t - 1 + \nu), \end{cases}$$

$$(3.28)$$

where $t \in \mathbb{N}_{a+1-\nu}$ and $0 < \nu < 1$. Actually, this fractional order form has recently been reported in [82]. By following the same steps as in

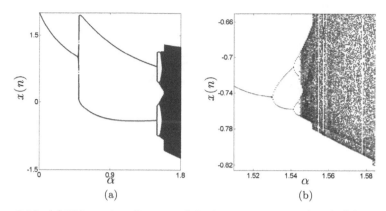

Figure 3.36. (a) Bifurcation diagram of the Hénon–Lozi map (3.26), (b) The local amplifications of bifurcation diagram for $\alpha \in (1.51, 1.583)$.

the previous subsections, we can then express the numerical formula of the fractional order Hénon–Lozi map (3.28) as follows:

$$
\begin{cases}
x(n) = x(0) + \dfrac{1}{\Gamma(\nu)} \displaystyle\sum_{j=1}^{n} \dfrac{\Gamma(n-j+\nu)}{\Gamma(n-j+1)} \\
\qquad \times (1 - \alpha S_\varepsilon \left(x \left(j-1\right)\right) + y(j-1) - x(j-1)), \\
y(n) = y(0) + \dfrac{1}{\Gamma(\nu)} \displaystyle\sum_{j=1}^{n} \dfrac{\Gamma(n-j+\nu)}{\Gamma(n-j+1)} (\beta x(j-1) - y(j-1)),
\end{cases}
$$

$$(3.29)$$

where $x(0)$ and $y(0)$ are the initial conditions. In fact, the above numerical formula allows us to examine the sensitivity of the fractional order Hénon–Lozi map (3.28) throughout the remainder of this section. To this aim and to fully understand the properties of such maps, some simulation experiments, including computing the largest Lyapunov exponents, constructing bifurcation diagrams, sketching phase portraits, and implementing the 0–1 test scheme, are preformed. We first look at the bifurcation graphs and give rough experimental bounds on the fractional order value to separate the asymptotically stable and chaotic ranges. In general, all performed numerical simulations reveal that the fractional order

Hénon–Lozi map have several coexistence attractors. In addition, the regions of the fractional order spaces corresponding to these coexisting attractors are illustrated using the bifurcation diagrams that can be computed based on the forward and backward construction strategies. However, a detailed investigation of the above discussion is presented in the next subsections.

3.5.1.1. *Bifurcations and largest Lyapunov exponents*

The bifurcation diagram and the largest Lyapunov exponents are deemed effective tools in identifying the complex dynamics of nonlinear systems (3.28). Typically, the fractional order Hénon–Lozi map displays a sensitive dependence on the initial conditions, where any two trajectories, which start from infinitesimally close points to those initial conditions, can diverge exponentially at the rate given by the largest Lyapunov exponent. The Lyapunov exponents can then be approximated using the Jacobian matrix J_i, which can be outlined in the fractional order Hénon–Lozi map (3.28) as:

$$J_i = \begin{pmatrix} a_i & b_i \\ c_i & d_i \end{pmatrix},$$

where

$$a_i = a_0 + \frac{1}{\Gamma(\nu)} \sum_{j=1}^{i} \frac{\Gamma(i-j+\nu)}{\Gamma(i-j+1)} \left(a_i \left(-\alpha \frac{\delta S_\varepsilon}{\delta x}(x) - 1 \right) + c_i \right),$$

$$b_i = b_0 + \frac{1}{\Gamma(\nu)} \sum_{j=1}^{i} \frac{\Gamma(i-j+\nu)}{\Gamma(i-j+1)} \left(b_i \left(-\alpha \frac{\delta S_\varepsilon}{\delta x}(x) - 1 \right) + d_i \right),$$

$$c_i = c_0 + \frac{1}{\Gamma(\nu)} \sum_{j=1}^{i} \frac{\Gamma(i-j+\nu)}{\Gamma(i-j+1)} (\beta a_i - c_i),$$

$$d_i = d_0 + \frac{1}{\Gamma(\nu)} \sum_{j=1}^{i} \frac{\Gamma(i-j+\nu)}{\Gamma(i-j+1)} (\beta b_i - d_i),$$

and where

$$\frac{\delta S_\varepsilon}{\delta x}(x) = \begin{cases} 1 & \text{if } x \geq \varepsilon, \\ \dfrac{x}{\varepsilon} & \text{if } |x| \leq \varepsilon. \\ -1 & \text{if } x \leq -\varepsilon. \end{cases} \tag{3.30}$$

By continuously changing the system's parameters, the fractional order map (3.28) can then undertake different dynamic scenarios. As can be seen in the standard case shown in Figure 3.36, the Hénon–Lozi map enters chaos via period-doubling bifurcation. Hence, it is interesting to visualize the effect of the fractional order value ν on the dynamic behavior when α is varied. The bifurcation diagram and the largest Lyapunov exponent in the $x\alpha$-plane are shown in Figure 3.37, where the parameters β and ε are set as $\varepsilon = 0.1$ and $\beta = 0.5$, while the value of ν is taken to be $\nu = 0.95$. It should be observed that when $\nu = 0.95$, the bifurcation diagram will be similar to that observed in Figure 3.36. In addition, we note that as ν passes into the range $(1.556, 1.74)$, the largest Lyapunov exponent becomes positive, which confirms the existence of chaos. In the meantime, the phase portraits of the fractional order Hénon–Lozi map (3.28)

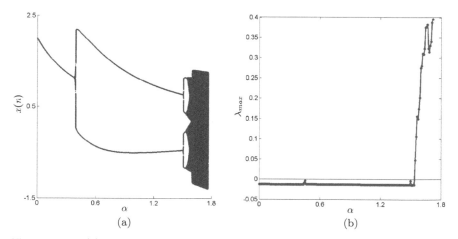

Figure 3.37. (a) Bifurcation diagram of the fractional order Hénon–Lozi map (3.28) versus α with $\nu = 0.95$. (b) The largest Lyapunov exponent of the fractional order map corresponding to (a).

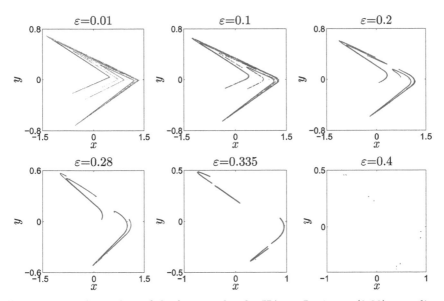

Figure 3.38. Phase plots of the fractional order Hénon–Lozi map (3.28) according to different values of ε.

for different values of ε with $\alpha = 1.7$, $\beta = 0.5$, and fractional order value $\nu = 0.95$ are shown in Figure 3.38. One could observe that the two states $x(n)$ and $y(n)$ approach a bounded attractor in all six considered cases. Moreover, the corresponding bounded attractors resemble those obtained in Figure 3.35.

In the same context, if the value of ν is reduced to 0.84, the resultant bifurcation diagram and the largest Lyapunov exponent will then be generated as shown in Figures 3.39(a)–(c). The local amplifications of the bifurcation diagram for $\alpha \in (1.4, 1.640)$ can also be shown in Figure 3.39(b). Regardless the value of ν, the period-doubling route to chaos is always observed. In addition, we observe that as α is increased, the fractional order Hénon–Lozi map (3.28) turns into chaos along with certain periodic orbits at $\alpha \simeq 1.62$. Besides, it is obviously seen that when the fractional order value ν decreases, the bifurcation diagram will be shifted gradually to the left, which indicates that the fractional order value ν represents indeed another bifurcation parameter.

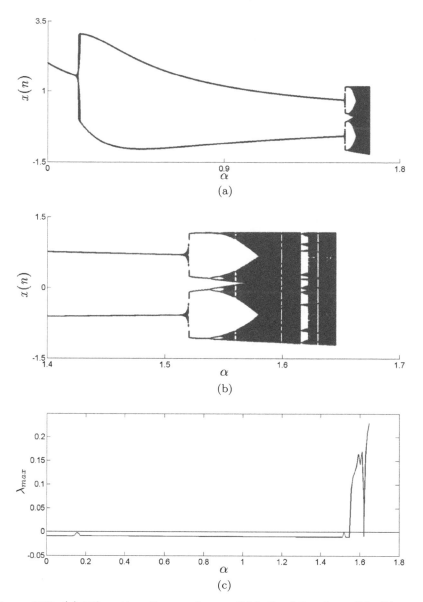

Figure 3.39. (a) Bifurcation diagram for $\nu = 0.84$, $\beta = 0.5$ and $\varepsilon = 0.1$; (b) zoom in on the bifurcation diagram; (c) the largest Lyapunov exponent.

To get further information about the complex dynamics, we plot some bifurcation diagrams of the fractional order Hénon–Lozi map versus ν, where the other parameters are taken as $\alpha = 1.7$ and $\beta = 0.6$. These bifurcation diagrams are obtained by presenting local maxima of the state $x(n)$ in terms of fractional order value ν, which is increased or decreased. In particular, Figures 3.40(a)–(c) display the bifurcation diagrams for $\varepsilon = 0.2$, $\varepsilon = 0.3$ and $\varepsilon = 0.335$, respectively. The coexisting bifurcation diagrams are plotted in blue and red colors. The trajectories that are colored red are generated in accordance with the gradual increase of the value of ν, while the trajectories colored blue are generated in accordance with the gradual decrease of the value of ν itself. Actually, this undoubtedly gives a better understanding of the effect of ν on the map's dynamics. In other words, it shows that the fractional order Hénon–Lozi map (3.28) experiences periodic and chaotic states according to the variation of the fractional order value ν. It is also interesting that the fractional order Hénon–Lozi map (3.28) has coexisting chaotic and periodic attractors. For instance, if we increase the value of ε to 0.335, the fractional order Hénon–Lozi map will then exhibit two different types of chaotic and periodic attractors for $\nu \in [0.964, 0.9712]$. In addition, it is worth noting that when $\varepsilon = 0.335$, the fractional order Hénon–Lozi map will generate chaotic behavior, while its integer order counterpart will generate a periodic motion. However, the coexisting attractors that are generated according to different values of ν and ε are shown in Figure 3.41.

3.5.1.2. *The 0–1 test method*

In order to further confirm the influence of the fractional order value ν on the properties of the fractional order Hénon–Lozi map, we apply here the 0–1 test method. For this purpose, we consider the translation components (p, q) of system (3.28) with two different values $\alpha = 1.65$, $\alpha = 1.6$, and with the fractional order value $\nu = 0.95$. Nevertheless, the (p, q)-trajectories can be shown in Figure 3.42. Obviously, these trajectories show a Brownian behavior

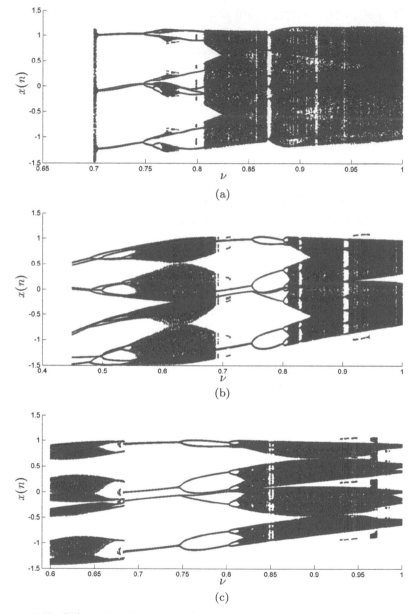

Figure 3.40. Bifurcation diagrams when ν is increased and when $\alpha = 1.7$ and $\beta = 0.5$ are decreased: (a) $\varepsilon = 0.2$, (b) $\varepsilon = 0.3$, (c) $\varepsilon = 0.335$.

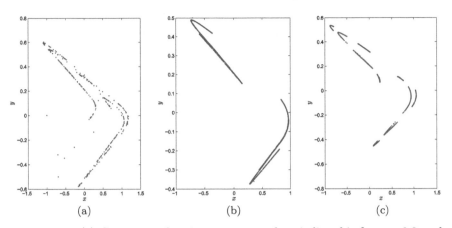

(a)　　　　　　　　(b)　　　　　　　　(c)

Figure 3.41. (a) Coexisting chaotic attractor and periodic orbit for $\varepsilon = 0.2$ and for the fractional order value $\nu = 0.9418$. (b) Chaotic attractors for $\varepsilon = 0.3$ and for the fractional order value $\nu = 0.9774$. (c) Coexisting chaotic attractors for $\varepsilon = 0.335$ and for the fractional order value $\nu = 0.9688$.

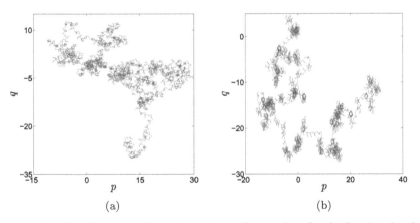

(a)　　　　　　　　　　　(b)

Figure 3.42. The Brownian-like trajectories in the pq-plane for the fractional order map (3.28) when $\nu = 0.95$, for (a) $\alpha = 1.65$, (b) $\alpha = 1.6$.

in the pq-plane. On the other hand, the asymptotic growth rate corresponding to Figure 3.42 converges to 1 as $n \longrightarrow \infty$ as in Figure 3.43. Based on these results, we have hence proved that the attractors for the two different values $\alpha = 1.65$ and $\alpha = 1.6$ are indeed chaotic.

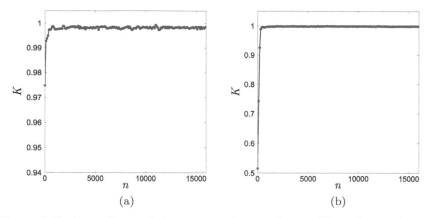

Figure 3.43. Dependence of the asymptotic growth rate K on the number of iterations n for the fractional order Hénon–Lozi map.

3.5.2. *Fractional Order Zeraoulia–Sprott Map*

Another interesting two-dimensional chaotic map that can be found in literature is the Zeraoulia–Sprott map [83]. This map has the following form:

$$\begin{cases} x\left(n+1\right) = \dfrac{-\alpha x\left(n\right)}{1+y^2\left(n\right)}, \\ y\left(n+1\right) = x\left(n\right) + \beta y\left(n\right), \end{cases} \tag{3.31}$$

where α and β are the bifurcation parameters. Actually, this map is considered to be a rich dynamic map as it can produce several chaotic attractors via the quasi periodic route to chaos. To see this, we consider the parameter values $\alpha = 3.7$ and $\beta = 0.6$. This, consequently, generates the chaotic attractor which can be depicted in Figure 3.44.

In order to continue with the investigation of the considered map, the fractional order version of this map is proposed here as follows:

$$\begin{cases} {}^{C}\Delta_a^\nu x\left(t\right) = \dfrac{-\alpha x\left(t-1+\nu\right)}{1+y^2\left(t-1+\nu\right)} - x\left(t-1+\nu\right), \\ {}^{C}\Delta_a^\nu y\left(t\right) = x\left(t-1+\nu\right) + \left(\beta-1\right) y\left(t-1+\nu\right), \end{cases} \tag{3.32}$$

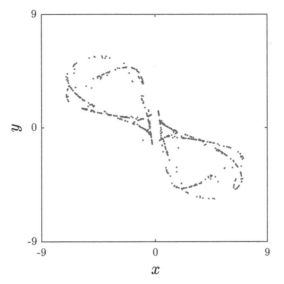

Figure 3.44. Phase portraits of the standard Zeraoulia–Sprott map with $\alpha = 3.7$ and $\beta = 0.6$.

for $t \in N_{a+1-\nu}$. Accordingly, the corresponding numerical formula of the above map can be described as:

$$
\begin{cases}
x\,(n) = x\,(0) + \dfrac{1}{\Gamma\,(\nu)} \sum_{j=1}^{n} \dfrac{\Gamma\,(n - j + \nu)}{\Gamma\,(n - j + 1)} \left(\dfrac{-\alpha x\,(j-1)}{1 + y^2\,(j-1)} - x\,(j-1) \right), \\[4mm]
y\,(n) = y\,(0) + \dfrac{1}{\Gamma\,(\nu)} \sum_{j=1}^{n} \dfrac{\Gamma\,(n - j + \nu)}{\Gamma\,(n - j + 1)} \left(x\,(j-1) + (\beta - 1)\,y\,(j-1) \right).
\end{cases}
$$

$$(3.33)$$

In order to assess the dynamics of this fractional order map, we intend to use certain numerical simulations. For attaining this goal, we will consider the initial conditions as $(x\,(0), y\,(0)) = (-1, -1)$, and we will also continuously vary the fractional order value ν. Figures 3.45 and 3.46 summarize the generated numerical results. In particular, the phase portraits of the fractional order map (3.32) for $\nu = 0.98, 0.95, 0.93, 0.92$ are indicated in Figure 3.45 with $\alpha = 3.7$ and $\beta = 0.6$. Besides, by taking the same value of β and with $\alpha \in [-1, 4]$, Figure 3.46 depicts the bifurcation diagrams for the four previous

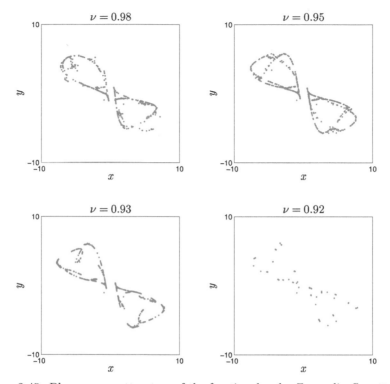

Figure 3.45. Phase space attractors of the fractional order Zeraoulia–Sprott map according to different fractional order values.

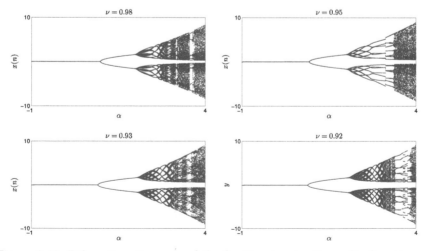

Figure 3.46. Bifurcation diagrams of the fractional order Zeraoulia–Sprott map according to different fractional order values.

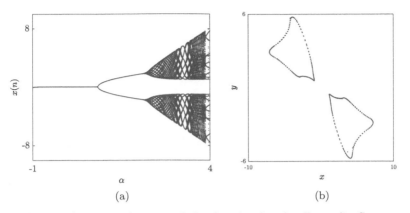

Figure 3.47. Bifurcation diagram of the fractional order Zeraoulia–Sprott map (left) and the attractor of the map's quasi periodic orbit for $\nu = 0.88$ (right).

values of the fractional order ν. It is noticed that system (3.32) exhibits chaotic behavior when $0.93 \leq \nu \leq 1$. In addition, we notice that as ν takes a value less than 0.93, chaos starts to disappear. We also note that when $0.8 \leq \nu \leq 0.88$, the map converges to a quasi periodic orbit as demonstrated in Figure 3.47.

Chapter 4

Chaos in 3D Discrete Fractional Systems

4.1. Introduction

In recent decades, the study carried out on chaos of three-dimensional discrete-time maps has attracted many researchers as it is considered as one of the most fascinating subjects exploring nonlinear dynamics. In particular, these maps have demonstrated a high degree of complexity when compared to others; one- or two-dimensional ones. This, actually, makes their constructions suitable in several real-life implementations such as pseudo-random number generation and image encryption. In the previous chapter, it was demonstrated that the two-dimensional fractional order maps offer the capability to reflect on the evolution of a practical problem and has more degrees of freedom than traditional discrete-time maps. To explore these notions further, we present many recent three-dimensional fractional order maps with hidden chaotic dynamics of equilibrium points. Such maps will be analyzed in relation to their dynamics via some numerical tools e.g. Lyapunov exponents, phase plots, bifurcation diagrams, and 0–1 test. The complexity of the ending map will be additionally investigated using the measures of ApEn. This leads to exploring the existence of chaos in these maps corresponding to different fractional order values.

4.2. Fractional Generalized Hénon Map

The integer order discrete-time system called the generalized Hénon map was proposed by Zheng *et al.* in [84]. Actually, this map is considered to be a generalization of the two-dimensional integer order Hénon map, and can be described as:

$$\begin{cases} x\,(n+1) = A - y^2\,(n) + Bz\,(n), \\ y\,(n+1) = x\,(n), \\ z\,(n+1) = y\,(n), \end{cases} \tag{4.1}$$

where A and B are the bifurcation parameters. As a matter of fact, this map exhibits chaos, especially when $(a, b) = (0.7281, 0.5)$ and $[x(0), y(0), z(0)] = [1, 0, 0]$, as demonstrated by the phase portraits shown in Figure 4.1. It is always helpful to examine the bifurcation diagram corresponding to a specific critical parameter. This would actually provide a comprehensive understanding of the dynamics of a chaotic system, see Figure 4.2. The integer order discrete-time generalized Hénon map (4.1) can be rewritten in the first-order difference form, i.e.

$$\begin{cases} \Delta x\,(n) = A - y^2\,(n) + Bz\,(n) - x(n), \\ \Delta y\,(n) = x\,(n) - y(n), \\ \Delta z\,(n) = y\,(n) - z(n). \end{cases} \tag{4.2}$$

In order to establish the fractional order discrete-time generalized Hénon map from (4.2), we operate the Caputo-like delta difference

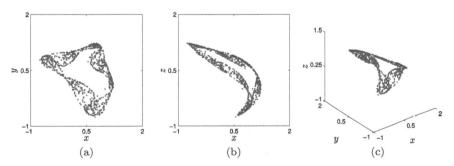

Figure 4.1. The chaotic attractors of map (4.1) in: (a) xy-plane (b) xz-plane (c) xyz-space.

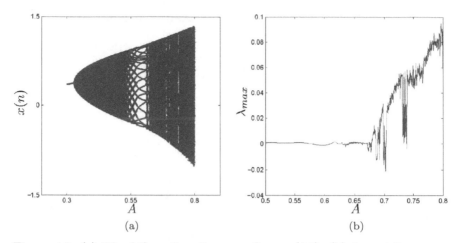

Figure 4.2. (a) The bifurcation diagram of map (4.1); (b) largest Lyapunov exponent.

operator on system (4.1). This gives the following form:

$$
\begin{cases}
{}^{C}\Delta_a^\nu x(t) = A - y^2(t-1+\nu) + Bz(t-1+\nu) - x(t-1+\nu), \\
{}^{C}\Delta_a^\nu y(t) = x(t-1+\nu) - y(t-1+\nu), \\
{}^{C}\Delta_a^\nu z(t) = y(t-1+\nu) - z(t-1+\nu),
\end{cases}
\tag{4.3}
$$

where $t \in N_{a+1-\nu}$ and $0 < \nu \le 1$. By setting $a = 0$, the following explicit numerical formulas are consequently yielded:

$$
\begin{cases}
x(n) = x(0) + \dfrac{1}{\Gamma(\nu)} \displaystyle\sum_{j=1}^{n} \dfrac{\Gamma(n-j+\nu)}{\Gamma(n-j+1)} \\
\qquad\quad \times \left(A - y^2(j-1) + Bz(j-1) - x(j-1)\right), \\[2mm]
y(n) = y(0) + \dfrac{1}{\Gamma(\nu)} \displaystyle\sum_{j=1}^{n} \dfrac{\Gamma(n-j+\nu)}{\Gamma(n-j+1)}(x(j-1) - y(j-1)), \\[2mm]
z(n) = z(0) + \dfrac{1}{\Gamma(\nu)} \displaystyle\sum_{j=1}^{n} \dfrac{\Gamma(n-j+\nu)}{\Gamma(n-j+1)}(y(j-1) - z(j-1)),
\end{cases}
\tag{4.4}
$$

where $x(0)$, $y(0)$ and $z(0)$ are the initial conditions.

By using the same initial conditions and the same bifurcation parameter values adopted for generating Figure 4.1, specific computer simulations are used to evaluate the numerical formulas (4.4) to gain a certain perspective on the dynamics of 3D fractional-order generalized Hénon map (4.4). The bifurcation diagram and its corresponding largest Lyapunov exponent for $\nu \in [0.96, 1]$ are illustrated in Figure 4.3. In Figure 4.3(a), we set $n = 700$ and plot only the last 200, then we compute the largest Lyapunov exponent using the Jacobian matrix algorithm. This figure actually visualizes how the fractional order value ν can effect the system behavior. In particular, we note that when $0 \leq \nu \leq 0.969$, the fractional-order map (4.4) diverges to infinity. On the contrary, it can be observed that there are vertical lines with a positive largest Lyapunov exponent when $\nu \in]0.969, 0.97[$. In this case, the solution $x(n)$ converges to a chaotic attractor. From Figures 4.3(a) and (b), we see that there is a transition from chaos to periodic cycles, followed by a series of appearance and disappearance of chaos. In other words, the largest Lyapunov exponent changes its values between negative numbers and positive over $\nu \in (0.97, 0.986)$. In addition, when $\nu \in [0.986, 1]$, the solution $x(n)$ always settles into a strange attractor.

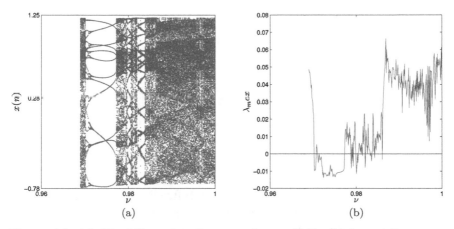

Figure 4.3. (a) The bifurcation diagram of map (4.1); (b) largest Lyapunov exponent.

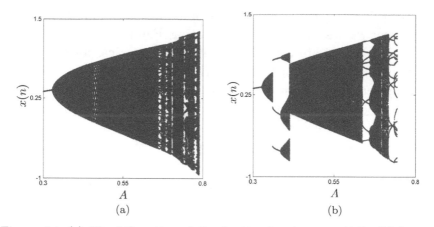

Figure 4.4. (a) The bifurcation of the fractional order map (4.3); (b) largest Lyapunov exponent with fractional order value ν.

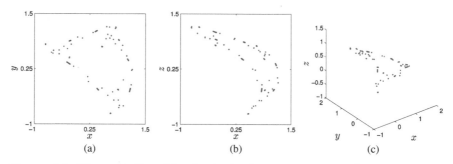

Figure 4.5. The periodic orbit obtained when $n = 2000$ and $\nu = 0.975$ in: (a) xy-plane; (b) xz-plane; (c) (x, y, z)-space.

To illustrate further, the bifurcation diagrams for $A \in [0.3, 0.8]$ are observed in Figure 4.4. In particular, in order to provide clearer diagrams, we set $n = 2000$, fix $B = 0.5$, and then discard away the first 1700 results. This provides the last 300 points as seen in Figure 4.5 corresponding to the fractional order values $\nu = 0.987$ and $\nu = 0.975$, respectively.

Similarly, Figure 4.4(a) shows the bifurcation diagram for $\nu = 0.987$. Clearly, based on such figures, we notice that when A passes to the interval $[0.66, 0.7885]$, a periodic cycle and a chaotic region are apparent. However, when the value of ν is set to be equal to 0.975, one notices the jumping behavior from chaotic set to a 3-periodic orbit, which suddenly drops into three small-sized attractors at $A = 0.326$. Besides, with the increase of parameter A, the fractional-order map (4.3) goes directly to a fully developed chaotic regime. It is worth pointing out here that this map never produces the same largest Lyapunov exponent twice, which confirms that each value of ν has its own attractor. At the same time, two chaotic attractors of the fractional order map (4.3), which include periodic orbit, are plotted according to $\nu = 0.987$ and for $\nu = 0.9695$, respectively in Figures 4.6 and 4.7.

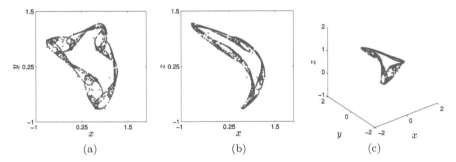

Figure 4.6. Chaotic attractor obtained when $n = 2000$ and $\nu = 0.987$ in: (a) xy-plane; (b) xz-plane; (c) (x, y, z)-space.

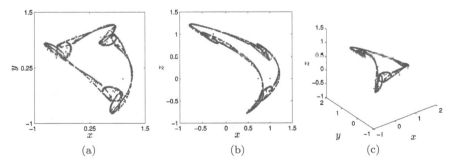

Figure 4.7. The periodic orbit obtained when $n = 2000$ and $\nu = 0.9695$ in: (a) xy-plane; (b) xz-plane; (c) (x, y, z)-space.

4.3. Fractional Generalized Hénon Map with Lorenz-Like Attractors

One of the most studied models in nonlinear systems is the two-dimensional Hénon map. The generalization of the classical version of this map is the three-dimensional Hénon map, which can be described by the following system of difference equations:

$$\begin{cases} x\,(n+1) = M_1 + Bz\,(n) + M_2 y\,(n) - x^2\,(n), \\ y\,(n+1) = x\,(n), \\ z\,(n+1) = y\,(n), \end{cases} \quad (4.5)$$

where B, M_1 and M_2 are the system's parameters and n represents the discrete iteration step. In fact, the three-dimensional Hénon map (4.5) is a simple quadratic diffeomorphism with a constant Jacobian. This map can exhibit wild Lorenz-type strange attractors. Such kind of attractors persist for open domains in the parameter space. A more comprehensive study was performed in [85]. For instance, the bifurcation of map (4.5) subject to different scenarios and initial settings was studied in [86, 87]. In order to visualize the dynamics of this map, we resort to phase plots, bifurcation diagrams, evolution states and Lyapunov exponent estimation. This will be carried out by assuming the parameter values as $(B, M_1, M_2) = (0.7, 0, 0.85)$, and the initial states as $[x(0), y(0), z(0)] = [0, 0, 1]$. Accordingly, the phase plot of the three-dimensional Hénon map (4.5) can be depicted as in Figure 4.8. By fixing $M_1 = 0$, $M_2 = 0.85$ and varying B within

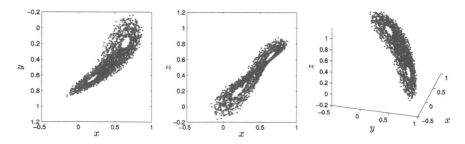

Figure 4.8. The phase plot of the three-dimensional Hénon map (4.5).

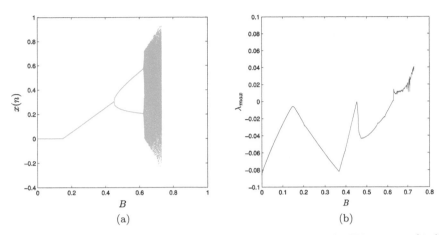

Figure 4.9. (a) The bifurcation diagram of the 3D integer-order Hénon map (4.5) and (b) the largest Lyapunov exponent.

the interval $[0, 1]$, we plot Figure 4.9(a) depicting the bifurcation diagrams of the considered map with respect to different parameters. Figure 4.9(b) shows the largest Lyapunov exponent corresponding to the bifurcation diagram. Based on these figures, we can see that the estimated Lyapunov exponents of map (4.5) are positive, and so this map can display chaotic behavior.

In the remainder of this section, the 3D fractional-order discrete-time Hénon map is introduced, then several numerical tools are carried out to investigate the dynamics of such a proposed map. In the first place, let us formulate the system of difference equations that represent the fractional order generalized Hénon map with Lorenz-like attractors, as:

$$\begin{cases} {}^{C}\Delta_a^{\nu} x\left(t\right) = M_1 + Bz(t-1+\nu) + M_2 y(t-1+\nu) \\ \qquad\qquad - x^2(t-1+\nu) - x(t-1+\nu), \\ {}^{C}\Delta_a^{\nu} y\left(t\right) = x(t-1+\nu) - y(t-1+\nu), \\ {}^{C}\Delta_a^{\nu} z\left(t\right) = y(t-1+\nu) - z(t-1+\nu), \end{cases} \tag{4.6}$$

where $t \in N_{a+1-\nu}$ and $\nu \in [0, 1]$ is the fractional order value. Actually, this map represents a generalization of the integer order Hénon map (4.5). However, in order to deal with the fractional order Hénon-like map (4.6) and to employ the numerical tools that are typically used to describe the dynamics of such a map, we should define the following discrete formulas:

$$
\begin{cases}
x(n) = x(0) + \dfrac{1}{\Gamma(\nu)} \displaystyle\sum_{j=1}^{n} \dfrac{\Gamma(n-j+\nu)}{\Gamma(n-j+1)} \\[2mm]
\qquad \times \left(M_1 + Bz(j-1) + M_2 y(j-1) - x^2(j-1) - x(j-1) \right), \\[3mm]
y(n) = y(0) + \dfrac{1}{\Gamma(\nu)} \displaystyle\sum_{j=1}^{n} \dfrac{\Gamma(n-j+\nu)}{\Gamma(n-j+1)} \left(x(j-1) - y(j-1) \right), \\[3mm]
z(n) = z(0) + \dfrac{1}{\Gamma(\nu)} \displaystyle\sum_{j=1}^{n} \dfrac{\Gamma(n-j+\nu)}{\Gamma(n-j+1)} \Gamma(n-j+1) \\[2mm]
\qquad \times \left(y(j-1) - z(j-1) \right).
\end{cases}
$$

$$(4.7)$$

In this part, with the help of the iterative formula (4.7), we are going to investigate the effect of the fractional order value ν on the behavior of the fractional order Hénon map (4.6).

4.3.1. *Bifurcation and Chaotic Attractors*

In order to continue exploring the dynamics of map (4.6), we set the system's parameters as $M_1 = 0$, $B = 0.7$, $M_2 = 0.85$, the initial conditions as $[x(0), y(0), z(0)] = [0, 0, 1]$, and also we let $n = 2500$. Herein, we consider the first 50 solutions as transient. For different values of ν, the numerical solutions of the fractional order Hénon map (4.6) are indicated in Figures 4.10–4.12. It is clear that when $\nu \geq 0.952$, the initial states $[x(0), y(0), z(0)] = [0, 0, 1]$ will converge to a kind of bounded attractors. Besides, Figures 4.10–4.12 illustrate that each value of ν can generate its own attractor. On the contrary, when ν is less then 0.952, the trajectories diverge to infinity; this is clearly shown in Figure 4.13. In fact, the latter figure shows the bifurcation diagram and the largest Lyapunov exponent diagram of $x(n)$ versus ν with respect to the given parameters M_1, M_2, B and initial conditions.

Now, if we fix $M_1 = 0$, $M_2 = 0.85$, and let B be varied from 0 to 1 with step size $\Delta B = 0.001$, then we can generate the

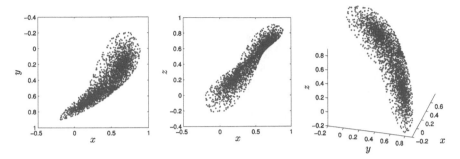

Figure 4.10. The phase plot of the fractional order Hénon map (4.6) when $\nu = 0.985$.

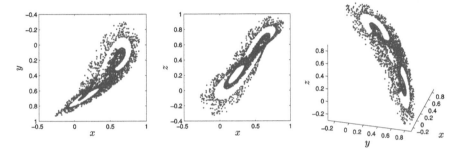

Figure 4.11. The phase plot of the fractional order Hénon map (4.6) when $\nu = 0.96$.

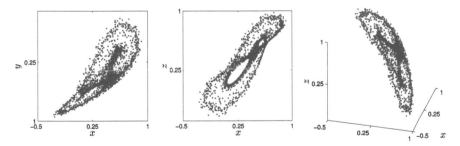

Figure 4.12. The phase plot of the fractional order Hénon map (4.6) when $\nu = 0.956$.

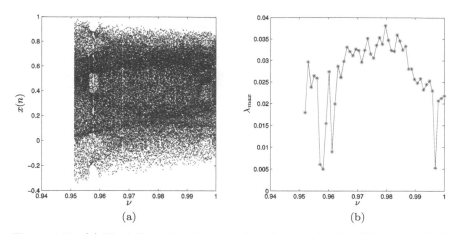

(a) (b)

Figure 4.13. (a) The bifurcation diagram of the fractional order Hénon map (4.6) as a function of ν, (b) largest Lyapunov exponent corresponding to (a).

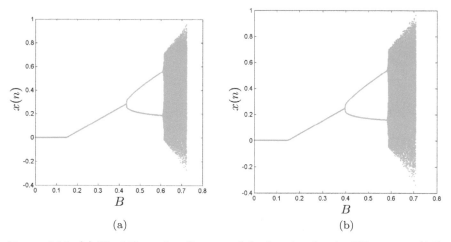

(a) (b)

Figure 4.14. (a) The bifurcation diagram of the fractional order Hénon map (4.6) as a function of B, when (a) $\nu = 0.985$ (b) $\nu = 0.96$.

bifurcation diagrams as in Figure 4.14 subject to the initial condition $[x(0), y(0), z(0)] = [0, 0, 1]$, and according to the fractional order values $\nu = 0.985$ and $\nu = 0.96$. Additionally, the dynamic behavior of the fractional order map (4.6) can also be illustrated by the largest

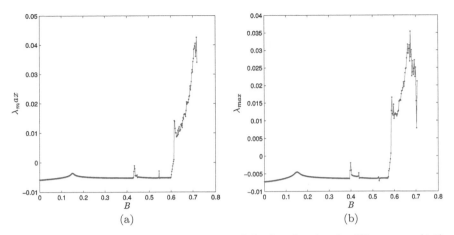

Figure 4.15. Largest Lyapunov exponent of the fractional order Hénon map (4.6) as a function of B, when (a) $\nu = 0.985$, (b) $\nu = 0.96$.

Lyapunov exponent as shown in Figure 4.15 for the same bifurcation parameters and with $\nu = 0.985$, $\nu = 0.96$.

4.3.2. The 0–1 Test Method

In order to further analyze the influence of the fractional order value ν on the properties of the fractional order map (4.6), we aim here to re-examine the dynamical behavior of its states by the 0–1 test. For this purpose, we choose the parameters of such map as $M_1 = 0$, $B = 0.7$ and $M_2 = 0.85$, as well as we choose the fractional order values as $\nu = 0.985$ and $\nu = 0.96$. Consequently, Figure 4.16 is generated which shows the asymptotic growth rate K of the fractional order map (4.6). It should be noted, for both values of ν, that the asymptotic growth rate K approaches 1 as n increases; which indicates the existence of chaos. This result agrees well with the bifurcation diagram and the largest Lyapunov exponent as shown in Figures 4.14 and 4.15.

From Figure 4.14, we observe that when ν decreases, the bifurcation diagram along the B-axis shrinks and the difference is then very small. But when ν hits the value 0.951, the unbounded area reaches the value $B = 0.7$, which confirms the above result.

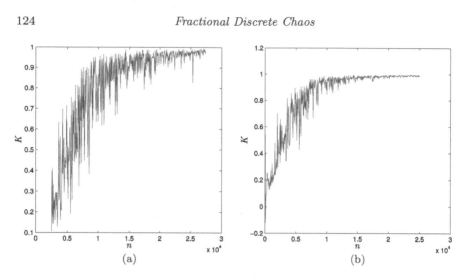

Figure 4.16. The 0–1 test for chaos, when (a) $\nu = 0.985$, (b) $\nu = 0.96$.

From the previous numerical results, it is seen clearly that chaos exists in the fractional order map (4.6) when ν lies within the range $(0.952, 1)$.

4.4. Fractional Stefanski Map

In [88], Stefanski introduced a system for generalization of the standard Hénon map into three-dimensional space. This system can be described as:

$$\begin{cases} x(n+1) = 1 + z(n) - \alpha y^2(n), \\ y(n+1) = 1 + \beta y(n) - \alpha x^2(n), \\ z(n+1) = \beta x(n), \end{cases} \quad (4.8)$$

where $x(n)$, $y(n)$, $z(n)$ are the states of the system, $\alpha > 0$, and $0 < \beta < 1$. In fact, this system has been studied extensively in the literature and is known to exhibit hyperchaotic behavior for the bifurcation parameters $\beta = 0.2$ and $\alpha \in [1.22, 1.40]$. The resulting attractor for $\alpha = 1.4$ and $\beta = 0.2$ is depicted in Figure 4.17.

With the help of discrete fractional calculus, we can operate the νth-Caputo formula for the integer order discrete-time system

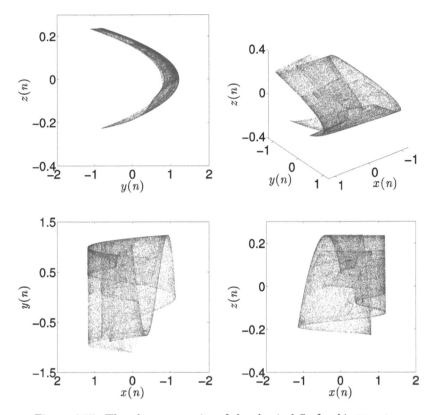

Figure 4.17. The phase portraits of the classical Stefanski attractor.

(4.8) over time $t \in \mathbb{N}_{a+1-\nu}$ and according to fractional order value $0 < \nu \leq 1$ to obtain:

$$
\begin{cases}
{}^{C}\Delta_{a}^{\nu} x\,(t) - 1 + z\,(t - 1 + \nu) - \alpha y^2\,(t - 1 + \nu) - x\,(t - 1 + \nu), \\
{}^{C}\Delta_{a}^{\nu} y\,(t) = 1 + (\beta - 1)\,y\,(t - 1 + \nu) - \alpha x^2\,(t - 1 + \nu), \\
{}^{C}\Delta_{a}^{\nu} z\,(t) = \beta x\,(t - 1 + \nu) - z\,(t - 1 + \nu).
\end{cases}
$$

$$(4.9)$$

Herein, we will consider system (4.9) as the fractional order Stefanski map. Before we go ahead and present some important dynamics of such a map, let us first present a discrete numerical formula equivalent to its formulation. For this purpose, we suppose that $a = 0$

to generate the following numerical formulas:

$$
\left\{
\begin{aligned}
x(n) &= x(0) + \frac{1}{\Gamma(\nu)} \sum_{j=1}^{n} \frac{\Gamma(n-j+\nu)}{\Gamma(n-j+1)} \\
&\quad \times \left(1 + z(j-1) - \alpha y^2(j-1) - x(j-1)\right), \\
y(n) &= y(0) + \frac{1}{\Gamma(\nu)} \sum_{j=1}^{n} \frac{\Gamma(n-j+\nu)}{\Gamma(n-j+1)} \\
&\quad \times \left(1 + (\beta-1)y(j-1) - \alpha x(j-1)\right), \\
z(n) &= z(0) + \frac{1}{\Gamma(\nu)} \sum_{j=1}^{n} \frac{\Gamma(n-j+\nu)}{\Gamma(n-j+1)} \left(\beta x(j-1) - z(j-1)\right),
\end{aligned}
\right.
$$

$$(4.10)$$

where $x(0)$, $y(0)$, and $z(0)$ are the initial conditions of the system.

In order to discuss the dynamical analysis of the fractional order Stefanski map (4.9), we will first let $a = 0$ and $x(0) = y(0) = z(0) = 0$ in such a system. For performing the numerical simulations, the step size $\Delta \alpha$ is chosen here to be equal to 0.001. Consequently, the phase portraits are then generated as in Figures 4.18 and 4.19 according to different fractional order values. The bifurcation diagrams are, at the same time, plotted in Figure 4.20 according to different fractional order values ν. Obviously, when $\nu = 0.97$, the bifurcation diagrams show a period-doubling cascade route to chaos in the range $\alpha \in [1.1, 1.4]$. Besides, as the value of ν decreases, the bifurcation diagram of the fractional order Stefanski map (4.9) expands along the α-axis and gradually shifts to the left.

In addition to visualizing the effect of parameter α on the dynamics of map (4.9), it should be noted that the fractional order value ν has another impact on its dynamics. This assertion can be further investigated by plotting again the bifurcation of the fractional order Stefanski map (4.9), but this time by taking the fractional order value ν as the critical parameter. The bifurcation diagram when $(\alpha, \beta) = (1.4, 0.2)$ and $[x(0), y(0)\, z(0)] = [0, 0, 0]$ can be seen as Figure 4.21. Clearly, we observe that the chaos is apparent over the

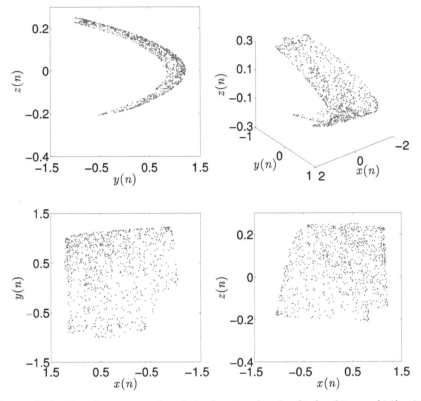

Figure 4.18. The phase portraits of the fractional order Stefanski map (4.9) with $\nu = 0.97$.

interval $\nu \in [0.915, 1]$. In addition to this, we also observe that as soon as ν drops below 0.915, the states diverge towards infinity.

Although the bifurcation plots clearly indicate the existence of chaos in the fractional order map at hand, it is usually more convenient to calculate or estimate its Lyapunov exponents. A common method to estimate Lyapunov exponents of the standard maps is to use a QR decomposition of the time-varying Jacobian matrix. For the fractional order maps, the Jacobian matrix can be calculated in a similar manner, and the Lyapunov exponents can be estimated. Figure 4.22 shows the estimated Lyapunov exponents according to different fractional order values ν, from which we observe that the results agree with those of bifurcation. In other words, lowering ν

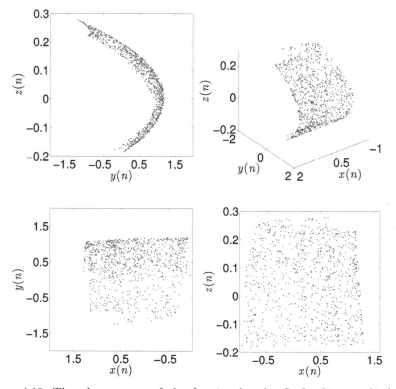

Figure 4.19. The phase space of the fractional order Stefanski map (4.9) for $\nu = 0.969$.

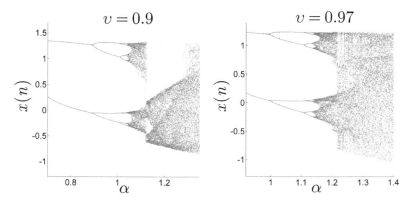

Figure 4.20. The bifurcation diagrams of the fractional order Stefanski map (4.9) with α as the critical parameter, $\beta = 0.2$, and according to different fractional order values ν.

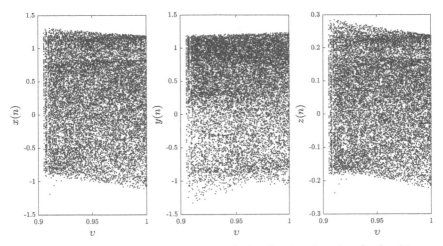

Figure 4.21. The bifurcation diagram of the fractional order Stefanski map (4.9) with $\nu \in [0,1]$ as the critical parameter, $(\alpha, \beta) = (1.4, 0.2)$, and $(x(0), y(0) z(0)) = (0, 0, 0)$.

below 1 yields lower exponents to the point that will later become negative, coinciding with a stable system.

4.5. Fractional Rössler System

In what follows, the three-dimensional integer order discrete-time Rössler map introduced in [68] is going to be considered. This map has the form:

$$\begin{cases} x(n+1) = b_1 x(n)(1 - x(n)) - b_2(z(n) + b_3)(1 - 2y(n)), \\ y(n+1) = b_4 y(n)(1 - y(n)) + b_5 z(n), \\ z(n+1) = b_6(1 - b_7 x(n))[(z(n) + b_3)(1 - 2y(n)) - 1], \end{cases}$$

(4.11)

where $x(n)$, $y(n)$, $z(n)$ are the states of the system with parameters $b_1 = 3.8$, $b_2 = 0.05$, $b_3 = 0.35$, $b_4 = 3.78$, $b_5 = 0.2$, $b_6 = 0.1$, and $b_7 = 1.9$. As a matter of fact, the integer order Rössler map is a well-known map in the field of difference systems as it has been extensively examined and applied in countless studies. The phase-space portraits of the integer-order Rössler map subject to the initial conditions $[x(0), y(0), z(0)] = [0.1, 0.2, -0.5]$ are seen in Figure 4.23.

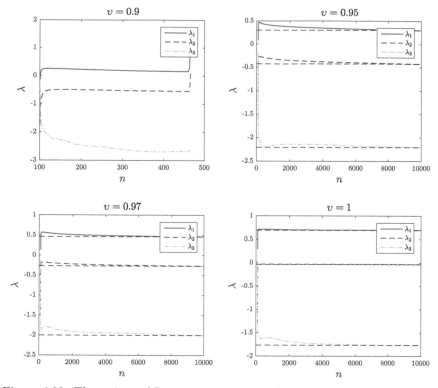

Figure 4.22. The estimated Lyapunov exponents of the fractional order Stefanski map (4.9) for $(\alpha, \beta) = (1.4, 0.2)$, $[x(0), y(0)\, z(0)] = [0, 0, 0]$, and according to different fractional order values.

In a similar manner to the fractional order Stefanski map, the fractional order map corresponding to map (4.11) can be described as

$$
\begin{cases}
{}^{C}\Delta_a^{\nu} x(t) = b_1 x(t - 1 + \nu)(1 - x(t - 1 + \nu)) \\
\qquad - b_2\left((z(t - 1 + \nu) + b_3)(1 - 2y(t - 1 + \nu))\right) \\
\qquad - x(t - 1 + \nu), \\
{}^{C}\Delta_a^{\nu} y(t) = b_4 y(t - 1 + \nu)(1 - y(t - 1 + \nu)) + b_5 z(t - 1 + \nu) \\
\qquad - y(t - 1 + \nu), \\
{}^{C}\Delta_a^{\nu} z(t) = b_6(1 - b_7 x(t - 1 + \nu)) \\
\qquad \times [(z(t - 1 + \nu) + b_3)(1 - 2y(t - 1 + \nu)) - 1] \\
\qquad - z(t - 1 + \nu),
\end{cases}
$$

$$(4.12)$$

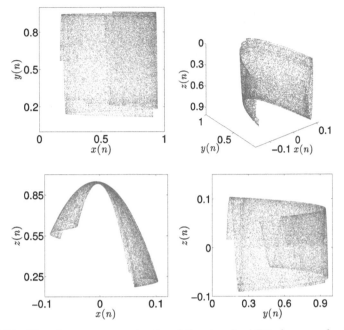

Figure 4.23. The phase-space portraits of the standard Rössler map for $b_1 = 3.8$, $b_2 = 0.05$, $b_3 = 0.35$, $b_4 = 3.78$, $b_5 = 0.2$, $b_6 = 0.1$, and $b_7 = 1.9$.

for $t \in \mathbb{N}_{a-v+1}$, where $0 < \nu \leq 1$. As a consequence of the aforesaid preamble, the corresponding numerical formula to system (4.12) can be expressed as:

$$
\begin{cases}
x(n) = x(a) + \dfrac{1}{\Gamma(\nu)} \displaystyle\sum_{j=1}^{n} \dfrac{\Gamma(n-j+\nu)}{\Gamma(n-j+1)} \\
\qquad \times \left(b_1 x(j-1)(1 - x(j-1)) - b_2(z(j-1) + b_3) \right. \\
\qquad \times \left. (1 - 2y(j-1)) - x(j-1)\right), \\
y(n) = y(a) + \dfrac{1}{\Gamma(\nu)} \displaystyle\sum_{j=1}^{n} \dfrac{\Gamma(n-j+\nu)}{\Gamma(n-j+1)} \\
\qquad \times \left(b_4 y(j-1)(1 - y(j-1)) + b_5 z(j-1) - y(j-1)\right), \\
z(n) = z(a) + \dfrac{1}{\Gamma(\nu)} \displaystyle\sum_{j=1}^{n} \dfrac{\Gamma(n-j+\nu)}{\Gamma(n-j+1)} \\
\qquad \times \left(b_6(1 - b_7 x(j-1))[(z(j-1) + b_3) \right. \\
\qquad \times \left. (1 - 2y(j-1)) - 1] - z(j-1)\right).
\end{cases}
$$

$$(4.13)$$

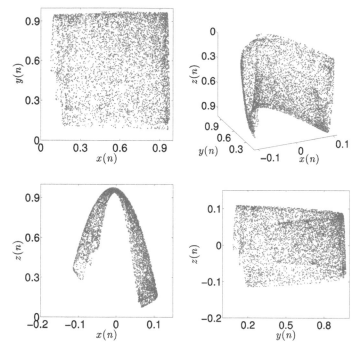

Figure 4.24. The phase portraits of the fractional order Rössler map (4.12) according to $\nu = 0.97$.

In order to ensure that the fractional order Rössler map (4.12) is chaotic, its bifurcation plot should be visualized. Hence, let us consider the simple case that assumes $a = 0$ and $[x(0), y(0), z(0)] = [0.1, 0.2, -0.5]$. As a result of this assumption, Figure 4.24 depicts the phase portraits of map (4.12) according to the fractional order value $\nu = 0.97$. Through an experimental sweep of the fractional order value ν, we found that the minimum value of ν for which system (4.12) exhibits a chaotic behavior is 0.903. Figure 4.25 shows the bifurcation diagram for parameter b_1, which is taken as the critical parameter, and according to $\nu = 0.97$ together with $(b_2, b_3, b_4, b_5, b_6, b_7) = (0.05, 0.35, 3.78, 0.2, 0.1, 1.9)$. This critical parameter is varied with step size $\Delta b_1 = 0.001$. At the same time, Figure 4.26 shows the bifurcation diagram of the fractional order Rössler map (4.12) for $\nu \in [0.9, 1]$ taken as the critical parameter, $(b_2, b_3, b_4, b_5, b_6, b_7) = (0.05, 0.35, 3.78, 0.2, 0.1, 1.9)$ and

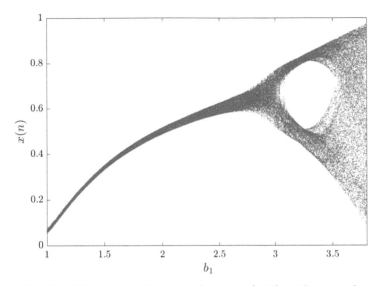

Figure 4.25. The bifurcation diagram of system (4.12) with a as the critical parameter and $\nu = 0.97$.

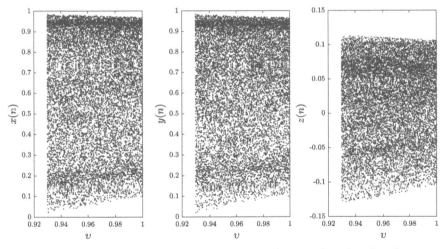

Figure 4.26. The bifurcation diagram of the fractional order Rössler map (4.12) with $\nu \in [0,1]$ as the critical parameter, $(b_2, b_3, b_4, b_5, b_6, b_7) = (0.05, 0.35, 3.78, 0.2, 0.1, 1.9)$, and $[x(0), y(0), z(0)] = [0.1, 0.2, -0.5]$.

$[x(0), y(0), z(0)] = [0.1, 0.2, -0.5]$. Based on these results, we observe that chaos is only observed for $\nu > \nu_0 \approx 0.933$. In addition, below ν_0, the considered map can become unstable and its states diverge towards infinity.

In order to estimate the Lyapunov exponents of system (4.12), we use the same parameters and initial conditions as reported above. As a result, we generate Figure 4.27 that shows those estimated exponents using the Jacobian matrix. For $\nu = 1$, we observe that $\lambda_1 \approx \lambda_2 > 0$, indicating a hyperchaotic nature of the fractional order Rössler map (4.12). Similar to the Stefanski map, we note that as ν reduces, the Lyapunov exponents do so too.

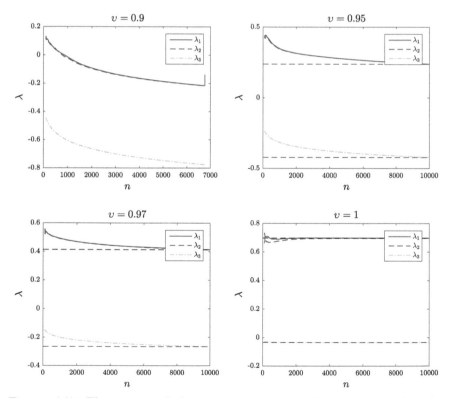

Figure 4.27. The estimated Lyapunov exponents of the fractional order Rössler map (4.12) for $(b_2, b_3, b_4, b_5, b_6, b_7) = (0.05, 0.35, 3.78, 0.2, 0.1, 1.9)$, $[x(0), y(0), z(0)] = [0.1, 0.2, -0.5]$, and according to different fractional order values ν.

4.6. Fractional Wang Map

In the following, another 3D chaotic map is presented called the integer order Wang map. This map, which was proposed in [89], has an interesting attractor, and can be described as

$$\begin{cases} x\,(n+1) = a_3 y\,(n) + (a_4+1)x\,(n), \\ y\,(n+1) = a_1 x\,(n) + y\,(n) + a_2 z\,(n), \\ z\,(n+1) = (a_7+1)z\,(n) + a_6 y\,(n)\,z\,(n) + a_5. \end{cases} \qquad (4.14)$$

Figure 4.28 shows the phase portraits of map (4.14) according to the parameters $(a_1, a_2, a_3, a_4, a_5, a_6, a_7) = (-1.9, 0.2, 0.5, -2.3, 2, -0.6, -1.9)$. It is easy to see that this system exhibits chaos, a result that has been reported and studied in the literature.

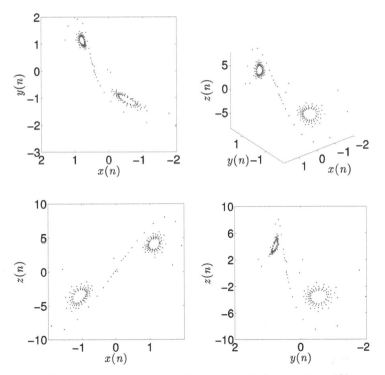

Figure 4.28. The phase portraits of the standard discrete-time Wang system (4.14).

Next, by following the same lines of the previous sections, we arrive at the fractional order discrete-time Wang map given as:

$$\begin{cases} ^{C}\Delta_a^{\nu} x\left(t\right) = a_3 y\left(t-1+\nu\right) + a_4 x\left(t-1+\nu\right), \\ ^{C}\Delta_a^{\nu} y\left(t\right) = a_1 x\left(t-1+\nu\right) + a_2 z\left(t-1+\nu\right), \\ ^{C}\Delta_a^{\nu} z\left(t\right) = a_7 z\left(t-1+\nu\right) + a_6 y\left(t-1+\nu\right) z\left(t-1+\nu\right) + a_5. \end{cases}$$
$$(4.15)$$

The numerical formulas corresponding to system (4.15) can then be obtained by Theorem 1.11. This would yield the following formulas:

$$\begin{cases} x\left(n\right) = x(0) + \dfrac{1}{\Gamma\left(\nu\right)} \sum_{j=1}^{n} \dfrac{\Gamma\left(n-j+\nu\right)}{\Gamma\left(n-j+1\right)} \left(a_3 y\left(j-1\right) + a_4 x\left(j-1\right)\right), \\ \\ y\left(n\right) = y(0) + \dfrac{1}{\Gamma\left(\nu\right)} \sum_{j=1}^{n} \dfrac{\Gamma\left(n-j+\nu\right)}{\Gamma\left(n-j+1\right)} \left(a_1 x\left(j-1\right) + a_2 z\left(j-1\right)\right), \\ \\ z\left(n\right) = z(0) + \dfrac{1}{\Gamma\left(\nu\right)} \sum_{j=1}^{n} \dfrac{\Gamma\left(n-j+\nu\right)}{\Gamma\left(n-j+1\right)} \\ \qquad\qquad \times \left(a_7 z\left(j-1\right) + a_6 y\left(j-1\right) z\left(j-1\right) + a_5\right). \end{cases}$$
$$(4.16)$$

In order to emphasize that the fractional order Wang map (4.15) is chaotic, we consider the case that assumes $a = 0$ and the initial conditions $[x\left(0\right), y\left(0\right), z\left(0\right)] = [0.05, 0.03, 0.02]$. Consequently, Figures 4.29 and 4.30 show several attractors generated from system (4.15) according to the two fractional order values $\nu = 0.97$ and $\nu = 0.969$. In addition, we have also plotted the bifurcation diagram with the critical parameter a_3 being varied at steps of $\Delta a_{.3} = 0.001$, and according to the remaining parameters chosen as $(a_1, a_2, a_4, a_5, a_6, a_7) = (-1.9, 0.2, -2.3, 2, -0.6, -1.9)$. However, the bifurcation duration is set to be $n = 200$. This generates bifurcation diagrams, as depicted in Figure 4.31 according to different fractional order values ν. At the same time, Figure 4.32 shows the bifurcation diagram of the fractional order Wang map (4.15) with $\nu \in [0.9, 1]$ as the critical parameter. Based on this figure, we notice that the map under consideration exhibits chaotic behavior over a short interval

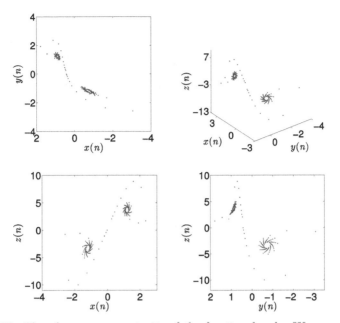

Figure 4.29. The phase space portraits of the fractional order Wang map (4.15) for $\nu = 0.97$.

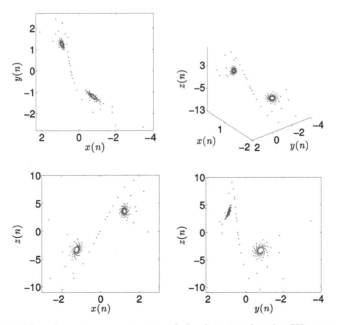

Figure 4.30. The phase-space portraits of the fractional order Wang map (4.15) for $\nu = 0.969$.

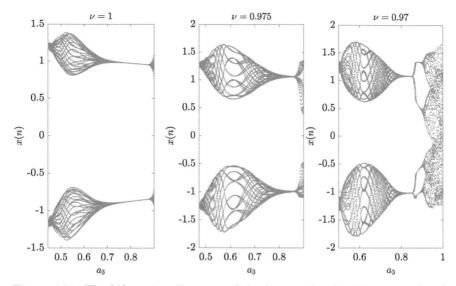

Figure 4.31. The bifurcation diagrams of the fractional order Wang map (4.15) according to different fractional order values.

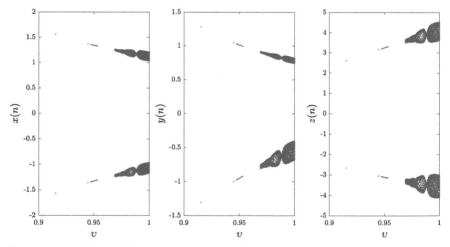

Figure 4.32. The bifurcation diagram of the fractional order Wang map (4.15) with $\nu \in [0,1]$ as the critical parameter, $(a_1, a_2, a_4, a_5, a_6, a_7) = (-1.9, 0.2, -2.3, 2, -0.6, -1.9)$, and $[x(0), y(0), z(0)] = [0.05, 0.03, 0.02]$.

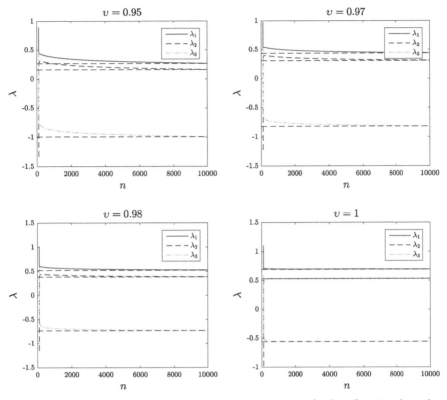

Figure 4.33. The estimated Lyapunov exponents of the fractional order Wang map (4.15) with $(a_1, a_2, a_4, a_5, a_6, a_7) = (-1.9, 0.2, -2.3, 2, -0.6, -1.9)$, $[x(0), y(0), z(0)] = [0.05, 0.03, 0.02]$, and according to different fractional order values ν.

of fractional order values. It should be noted that chaos disappears completely when $\nu < \nu_0 \approx 0.915$. While when $\nu < 0.968$, the chaotic behavior is intermittent and has a very short range.

The Lyapunov exponents of the fractional order Wang map (4.15) with the same previous parameters and initial conditions are depicted in Figure 4.33. For $\nu = 1$, we observe that $\lambda_1 > \lambda_2 > 0$, indicating hyperchaotic dynamics of map (4.15). These changes occur as the fractional order value is made smaller. On the other hand, when $\nu = 0.9$ and $0 > \lambda_1 > \lambda_2 > \lambda_3$, a stable dynamic is generated of map (4.15).

4.7. Fractional Grassi–Miller Map

Over the years, many integer-order chaotic maps have been proposed. These maps have found applications in numerous fields within science and engineering. Herein, we are interested in the Grassi–Miller map, which was proposed first by Miller and Grassi in [90] as a generalization of the Hénon map. The standard Grassi–Miller map can be described by the following three-dimensional discrete-time system:

$$\begin{cases} x(n+1) = 1 + z(n) - \alpha y^2(n), \\ y(n+1) = 1 + \beta y(n) - \alpha x^2(n), \\ z(n+1) = \beta x(n), \end{cases} \quad (4.17)$$

where $x(n)$, $y(n)$, $z(n)$ denote the states of the system and α, β are the system's parameters. As demonstrated in [90], this system can exhibit hyperchaotic behavior, especially when $(\alpha, \beta) = (1.76, 0.1)$ subject to the initial conditions $[x(0), y(0), z(0)] = [1, 0.1, 0]$. The plots of the attractors generated by this map can be seen in Figure 4.34. From this viewpoint, we will examine the fractional order version of the integer order discrete-time Grassi–Miller system (4.17), by phase portraits and bifurcation diagrams of the proposed system that can be used to highlight the ranges of parameters and the fractional-order values over which chaos is observed.

By operating the Caputo fractional order difference operator on the integer order discrete-time Grassi–Miller system (4.17), we can formulate the fractional order Grassi–Miller map, which would be

$$\begin{cases} {}^{C}\Delta_a^\nu x(t) = 1 + z(t-1+\nu) - \alpha y^2(t-1+\nu) - x(t-1+\nu), \\ {}^{C}\Delta_a^\nu y(t) = 1 + \beta y(t-1+\nu) - \alpha x^2(t-1+\nu) - y(t-1+\nu), \\ {}^{C}\Delta_a^\nu z(t) = \beta x(t-1+\nu) - z(t-1+\nu), \end{cases}$$

$$(4.18)$$

where $0 < \nu \leq 1$ is the fractional-order value, and a defines the starting point of the set $\mathbb{N}_{a+1-\nu}$. The numerical formulas that

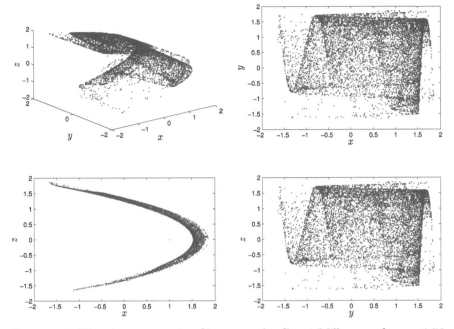

Figure 4.34. The phases portraits of integer order Grassi–Miller map for $\alpha = 1.76$ and $\beta = 0.1$.

correspond to map (4.18) can be outlined as follows:

$$
\begin{cases}
x(n) = x(a) + \dfrac{1}{\Gamma(\nu)} \sum_{j=1}^{n} \dfrac{\Gamma(n-j+\nu)}{\Gamma(n-j+\nu)} \\
\qquad \times \left(1 + z(j-1) - \alpha y^2(j-1) - x(j-1)\right), \\
y(n) = y(a) + \dfrac{1}{\Gamma(\nu)} \sum_{j=1}^{n} \dfrac{\Gamma(n-j+\nu)}{\Gamma(n-j+\nu)} \\
\qquad \times \left(1 + \beta y(j-1) - \alpha x(j-1) - y(j-1)\right), \\
z(n) = z(a) + \dfrac{1}{\Gamma(\nu)} \sum_{j=1}^{n} \dfrac{\Gamma(n-j+\nu)}{\Gamma(n-j+\nu)} \left(\beta x(j-1) - z(j-1)\right).
\end{cases}
$$

$$(4.19)$$

As a matter of fact, it is easy to get several simulations related to time series via computer programming as per the above equations.

Next, the value of a reported in (4.19) will be taken to be equal 0. Besides, the dynamics of map (4.18) will be, at the same time, analyzed.

In connection with the fractional order map (4.18), the experimental variation of the fractional order value has shown a variety of responses in the form of chaotic states, limit cycles and asymptotic stability. For observing the phase portraits, bifurcation diagrams, and maximum Lyapunov exponent of map (4.18), we use the numerical formula (4.19) through a MATLAB script along with different sets of parameters and fractional order values.

4.7.1. *Dynamics Analysis on Varying* α

In order to discuss the dynamics analysis of map (4.18) on varying α, we first let $\nu = 1$, $[x(0), y(0), z(0)] = [1, 0.1, 0]$, and $(\alpha, \beta) = (1.76, 0.1)$. As expected, when $\nu = 1$ is considered in the numerical formulas given in (4.19), the generated phase portraits obtained from map (4.18) are identical to those depicted in Figure 4.34 for map (4.17). Note that these simulations implemented for both systems are carried out when $n = 7000$.

Similarly, we consider next the case where $\nu = 0.98$. We plot Figures 4.35 and 4.36 that represent the portraits of map (4.18) so that $\alpha = 1.6$ and $\alpha = 1.5$, respectively. We notice that a slight change in system's parameters would lead to a variation in the strange attractor. The time evolution of the states of map (4.18) is depicted in Figure 4.38 for $\alpha = 1.6$. Figure 4.39 depicts the portraits of the same system when $\nu = 0.86$ and $\alpha = 1.5$. This confirms that the fractional order map (4.18) can exhibit chaotic behaviors for a range of different fractional order values and system's parameters. On the other hand, Figure 4.37 shows the phase portraits of map (4.18) subject to $\alpha = 0.4$ and $\nu = 0.5$. Obviously, in this scenario, the system does not exhibit chaos. This suggests that the existence or absence of chaos is dependent on the fractional order values.

In chaos theory, the phase portraits seem to be a nice tool for evaluating the nature of solutions, but they are not sufficient to gain a comprehensive perspective. So, we may also focus on bifurcation

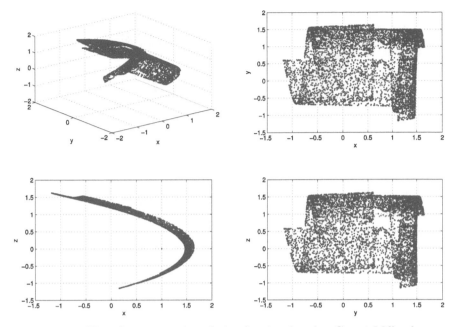

Figure 4.35. The phase portraits of the fractional order Grassi–Miller hyper-chaotic map (4.18) for $\beta = 0.1$, $\alpha = 1.6$, and $\nu = 0.98$.

plots, taking e.g. α as the critical parameter. We will also let the step size $\Delta\alpha$ be 0.002, and use the numerical formula (4.19). This will produce bifurcation diagrams for different fractional order values ν. For instance, when $\nu = 0.98$ and $\nu = 0.88$, the bifurcation diagrams are plotted in Figures 4.40(a) and (b), respectively. From these figures, it can be deduced that each time we modify the fractional order value, considerable changes occur in the chaotic region of the diagram. These observed variations in the shape of the bifurcation plot suggest that the fractional order Grassi–Miller map (4.18) is able to exhibit a variety of chaotic motions.

Let us now conduct a detailed analysis of the obtained plots to establish a connection between the changes in the shape of the attractors and the values of the system's parameters. If we compare the attractors outlined in the xy- and yz-plots as in Figure 4.34 (the integer order case with $\alpha = 1.76$) with the attractors outlined in the xy- and yz-plots as in Figure 4.35 (the fractional-order case

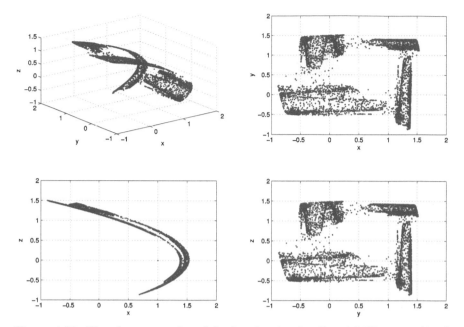

Figure 4.36. The phase portraits of the fractional order Grassi–Miller map (4.18) $\alpha = 1.5$ and $\nu = 0.98$.

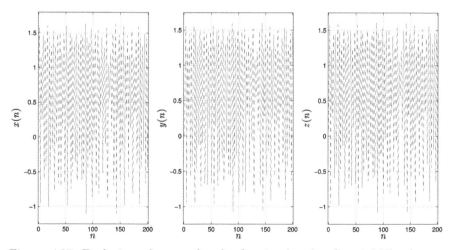

Figure 4.37. Evolution of states for the fractional order Grassi–Miller hyper-chaotic map (4.18) with $\alpha = 1.6$ and $\nu = 0.98$.

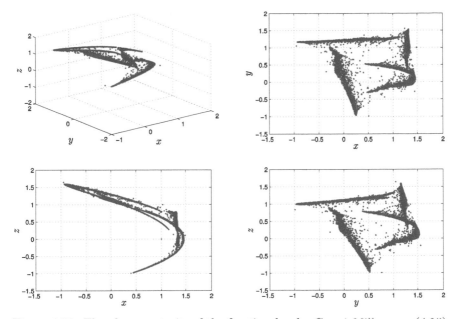

Figure 4.38. The phase portraits of the fractional order Grassi–Miller map (4.18) with $\alpha = 1.5$ and $\nu = 0.86$.

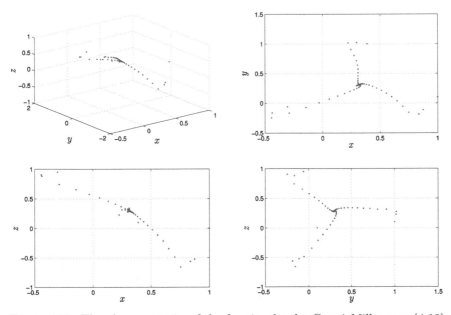

Figure 4.39. The phase portraits of the fractional order Grassi–Miller map (4.18) with $\alpha = 0.4$ and $\nu = 0.5$.

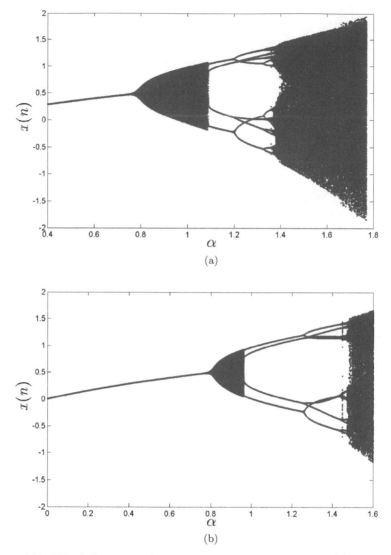

Figure 4.40. The bifurcation diagram versus α according to: (a) $\nu = 0.98$, (b) $\nu = 0.88$.

with $\nu = 0.98$ and $\alpha = 1.6$), we notice that a slight change in the parameter values (ν from 1 to 0.98 and α from 1.76 to 1.6) can lead to a reduction in the size of these attractors. In particular, in Figure 4.34, the attractors outlined in the xy- and yz-plots cover all the areas ranging from -2 to $+2$ for both the horizontal and the vertical axes, while in Figure 4.35 the attractors outlined in the xy- and yz-plots cover the area ranging approximately from -1.5 to $+1.5$ for both axes, indicating that the reduction of the system's parameters will generate a reduction in the size of attractors.

Next, we will discuss the results plotted in Figures 4.35 and 4.36. That is, if we compare the attractors outlined in the xy- and yz-plots shown in Figure 4.35 (the fractional order case when $\nu = 0.98$ and $\alpha = 1.6$) with the attractors outlined in the xy- and yz-plots shown in Figure 4.36 (the fractional order case when $\nu = 0.98$ and $\alpha = 1.5$), we notice that a slight change in the parameter values (α from 1.6 to 1.5) would generate a further reduction in the size of the attractor as well as its subdivision into four different parts. Specifically, if we refer to the size of the generated attractors, then we note that the attractors outlined in the xy- and yz-plots shown in Figure 4.36 cover the area ranging approximately from -1 to $+1.5$ for both axes, indicating that reducing the parameter's value α from 1.6 to 1.5 would generate a reduction in the size of those attractors (besides the appearance of four different parts that are related to the attractor plotted in Figure 4.35).

In order to go ahead with this analysis, we intend to discuss the results plotted in Figures 4.36 and 4.39. So, if we compare the attractors outlined in the xy- and yz-plots shown in Figure 4.36 (when $\nu = 0.98$ and $\alpha = 1.5$) with the attractors outlined in the xy- and yz-plots shown in Figure 4.38 (when $\nu = 0.86$ and $\alpha = 1.5$), we notice that the change in the fractional order value ν from 0.98 to 0.86 as well as keeping the same value of α would not generate a reduction in the size of the attractor anymore. However, it is noticed, based on Figure 4.39, that the shape of attractor is slightly different from that plotted in Figure 4.36, although the subdivision into four different parts can be seen in Figure 4.39. In addition, we note that when we further reduce the fractional order value of map (4.18), chaotic phenomena disappear as in Figure 4.37 for $\nu = 0.5$.

In the following, we intend to analyze the bifurcation diagrams with the aim of clarifying the values of the system's parameters for which the dynamics of the fractional order Grassi–Miller map (4.18) show asymptotic stability, limit cycles or chaotic behaviors. From Figure 4.40(a), we see the bifurcation diagram obtained for $\nu = 0.98$, it is clear that the map has a stable equilibrium point until the parameter α reaches the value of 0.7. Then, when the value of α is assumed between 0.8 and 0.9, the bifurcation diagram shows the typical shape that indicates the presence of chaotic behaviors. When the value $\alpha = 1$ is reached, the map is characterized by the presence of period-4 limit cycles. Successively, it is noticed that the map shows period-8 limit cycles when the value $\alpha = 1.3$ is reached. By further increasing the value of α, chaotic behaviors are generated for $\alpha = 1.5$ and beyond. Similar considerations can be done for the bifurcation diagram reported in Figure 4.40(b) for $\nu = 0.88$.

4.7.2. *Dynamics Analysis on Varying ν*

Although the above analysis gives a clear idea of the dynamics of the proposed fractional order map, a more comprehensive understanding of the map's dynamics may be obtained through visualizing the bifurcation diagrams and Lyapunov exponents as a function of the fractional order value ν. In this experiment, we keep the initial values as $[x(0), y(0), z(0)] = [1, 0.1, 0]$ and the parameter as $\beta = 0.1$, while we vary the fractional order value ν across the interval $[0, 1]$. As a result of these assumptions, we plot Figure 4.41 that shows the bifurcation diagrams and the largest Lyapunov exponent of map (4.18) for $\alpha \in \{1.6, 1.5, 1.3\}$. It should be noted when $\alpha = 1.6$, we obtain the results as in the top row of the figure. Besides, it should be observed that both diagrams show that when $0.92 \leq \nu \leq 1$, the fractional-order map (4.18) generates chaotic behavior, as the largest Lyapunov exponent remains positive. In addition, a transient region is observed over the interval $0.916 \leq \nu \leq 0.92$, then chaos is consequently observed again until it eventually disappears as $\nu \leq 0.814$ and the states of the system diverge towards infinity. Next, for $\alpha = 1.5$, the results are depicted in the middle row. We see that when $\nu \in [0.8, 1]$, the system exhibits chaotic behavior along with periodic windows. Herein, the lowest fractional order value that yields chaos is 0.8. Similarly, for $\alpha = 1.3$, the bifurcation diagram and the largest Lyapunov exponent for $\nu \in [0.75, 1]$ are depicted in

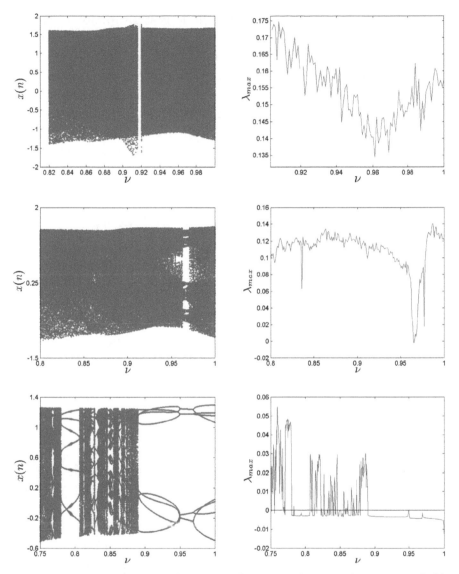

Figure 4.41. The bifurcation diagrams taking ν as the critical parameter (left) along with the corresponding largest Lyapunov exponents (right) for different parameter values: $(\alpha, \beta) = (1.6, 0.1)$ (top), $(\alpha, \beta) = (1.5, 0.1)$ (middle), and $(\alpha, \beta) = (1.3, 0.1)$ (bottom).

the bottom row. As it can be seen, the largest Lyapunov exponent
diagram matches perfectly with the bifurcation diagram. Moreover,
the fractional order Grassi–Miller map (4.18) remains in periodic
motion over the interval $[0.89, 1]$. However, when ν moves across the
interval $[0.75, 0.89)$, we notice that the largest Lyapunov exponents
alternate between positive and negative signs, indicating that the
periodic states become chaotic for certain values of ν. This analysis
tells us that the minimum fractional order value for which the chaotic
motion is observed depends on the system's parameters.

4.7.3. *The 0–1 Test*

To reflect more on the sensitivity of the fractional order map (4.18),
the 0–1 test is considered here. For this purpose, we plot Figures 4.42
and 4.43 depicting the results of the 0–1 test according to different
fractional order values of ν and the system's parameter α in which
$[x(0), y(0), z(0)] = [0.5, 0.1, 0]$. In particular, Figure 4.18 shows the
trajectories of the translation function of the fractional order Grassi–
Miller map (4.18) in the pq-plane for $\nu = 0.98$ and $\beta = 0.1$, by
varying α. Clearly, Figures 4.42(a)–(c) indicate bounded trajectories,
implying that map (4.18) is periodic. On the other hand, Figure
4.42(d) depicts the Brownian-like trajectories, indicating that map
(4.18) is chaotic when $\alpha = 1.55$, which confirms very well the results
shown in Figure 4.41.

Next, we apply the 0–1 test by taking $\alpha = 1.3$, $\beta = 0.1$, by
varying the fractional order value ν. The result can be seen in
Figure 4.43 that depicts the trajectories of the translation function
in pq-plane. In addition, Figures 4.43(a)–(b) show bounded-like
trajectories, indicating that the fractional order Grassi–Miller map
(4.18) is periodic when $\nu = 1$ and $\nu = 0.755$. Moreover, when
$\nu = 0.775$, the chaotic attractor is confirmed by the Brownian-like
trajectories as shown in Figure 4.43(c).

Based on the above numerical simulation, we notice that the
fractional order Grassi–Miller map (4.18) can exhibit chaos for a
range of parameters and fractional order values. It can be seen that
even with a fractional order value as low as $\nu = 0.5$, chaos still exists
for specific parameters. Generally, map (4.18) exhibits different types

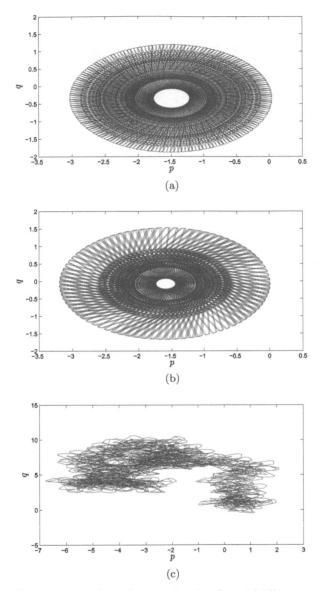

Figure 4.42. The 0–1 test of the fractional order Grassi–Miller map (4.18) with $\nu = 0.98$ and $\beta = 0.1$: (a) bounded-like trajectories for $\alpha = 0.7$, (b) bounded-like trajectories for $\alpha = 1$, (c) bounded-like trajectories for $\alpha = 1.375$, (d) Brownian-like trajectories for $\alpha = 1.55$.

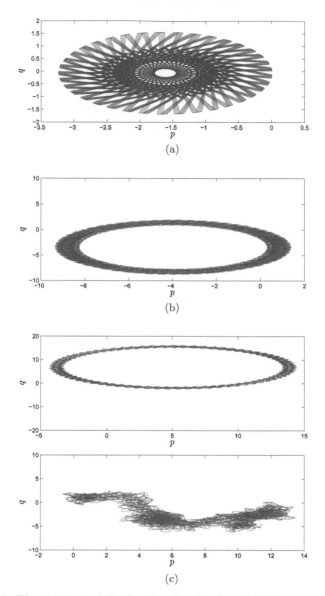

(a)

(b)

(c)

Figure 4.43. The 0–1 test of the fractional order Grassi–Miller map (4.18) with $\alpha = 1.3$ and $\beta = 0.1$: (a) bounded-like trajectories for $\nu = 1$, (b) bounded-like trajectories for $\nu = 0.755$, (c) Brownian-like trajectories for $\nu = 0.775$.

of attractors according to different fractional order values, that yield several chaotic regions. It is interesting that the system with integer order is periodic, but chaos is observed with the decrease of the fractional order value as well.

4.8. Fractional Cournot Game Model

Researchers' ever-growing interest in fractional calculus have led to the ever-broadening applications in many fields of science and engineering [91]. In economics, a number of mathematical models describing different phenomena have been introduced. Since the fractional order operators are nonlocal, they are suitable for constructing models characterized by memory effect. This is the reason why the fractional order difference systems perform better than that of integer order ones when economic phenomena are described over large periods of time [92].

More recently, attention has been focused on the presence of chaotic phenomena in economic systems described using integer order difference or differential equations. Unlike continuous-time systems, very few papers on chaotic phenomena described in economic systems by discrete-time dynamics have been published to date. These economic systems, which usually involve concepts from Cournot game theory applied to oligopolistic markets, generate complex dynamics that lead to bifurcations and chaos [93]. The behaviors of an oligopoly game are much complicated because firms must consider not only the market demand, but also the strategies of the competitors. The purpose of this section is to bring together two independent lines of research in applied mathematics and industrial economics by combining the fractional order difference equations with Cournot equilibria. For this purpose, we will reconsider the Cournot problem in the light of the theory of long memory. This will be implemented by proposing a fractional order discrete-time Cournot game model with three firms that produce differentiated products and compete over a common market. This would allow participants to make suitable decisions when they will make full use of their historical information.

4.8.1. *The Fractional Order Cournot Game Model with Long Memory*

For constructing the fractional order Cournot game model, we should consider a monopolistic market, where the three firms produce different products. The inverse demand function can be given as

$$p_i = \alpha - q_i - \beta \sum_{\substack{j=1 \\ j \neq i}}^{3} q_i, \quad i = 1, 2, 3, \tag{4.20}$$

where q_i denotes the outputs of products produced by the three firms, and the constants $\alpha > 0$, $0 < \beta < \sqrt{0.5}$ are the coefficients of the market demand function. So, we assume that the cost function of these firms is proposed in the following nonlinear form:

$$C_i(q_i) = \gamma_i q_i + \delta_i \sum_{\substack{j=1 \\ j \neq i}}^{3} q_i q_j, \quad i = 1, 2, 3, \tag{4.21}$$

where γ_i and δ_i are two positive constants. In this setting, the relative profit can be given by the difference between the absolute profit of the firm i and the sum of the other firms' profits. That is;

$$\Pi_i(q_1, q_2, q_3) = \left[q_i \left(\alpha - q_i - \beta \sum_{\substack{j=1 \\ j \neq i}}^{3} q_i \right) - \gamma_i q_i - \delta_i \sum_{\substack{j=1 \\ j \neq i}}^{3} q_i q_j \right]$$

$$- \sum_{\substack{j=1 \\ j \neq i}}^{3} p_j(q_1, q_2, q_3) q_j + C_j(q_j), \quad i = 1, 2, 3. \tag{4.22}$$

Through substituting Eqs. (4.20) and (4.21) into Eq. (4.22), the relative profit functions can then be given as:

$$\Pi_1(q_1, q_2, q_3) = \alpha(q_1 - q_2 - q_3) - q_1^2 + q_2^2 + q_3^2 + 2\beta q_2 q_3 - \gamma_1 q_1 + \gamma_2 q_2$$

$$+ \gamma_3 q_3 - \delta_1(q_1 q_2 + q_1 q_3) + \delta_2(q_1 q_2 + q_2 q_3)$$

$$+ \delta_3(q_1 q_3 + q_2 q_3),$$

$$\Pi_2(q_1, q_2, q_3) = \alpha(q_2 - q_1 - q_3) - q_2^2 + q_1^2 + q_3^2 + 2\beta q_1 q_3 - \gamma_2 q_2 + \gamma_1 q_1$$
$$+ \gamma_3 q_3 - \delta_2(q_1 q_2 + q_2 q_3) + \delta_1(q_1 q_2 + q_1 q_3)$$
$$+ \delta_3(q_1 q_3 + q_2 q_3),$$

$$\Pi_3(q_1, q_2, q_3) = \alpha(q_3 - q_1 - q_2) - q_3^2 + q_1^2 + q_2^2 + 2\beta q_1 q_2 - \gamma_2 q_3 + \gamma_1 q_1$$
$$+ \gamma_2 q_2 - \delta_3(q_1 q_3 + q_2 q_3) + \delta_1(q_1 q_2 + q_1 q_3)$$
$$+ \delta_2(q_1 q_2 + q_2 q_3). \tag{4.23}$$

With the above assumptions, the maximizing profit is obtained by setting $\frac{\partial \Pi_i}{\partial q_i} = 0$. In order to construct the integer order dynamical system of this game, we assume that each firm tries to use information based on the marginal profit $\frac{\partial \Pi_i}{\partial q_i}$. The mathematical model of this matter can be written as follows:

$$q_i(n+1) = q_i(n) + \varepsilon_i q_i(n) \frac{\partial \Pi_i}{\partial q_i}, \quad i = 1, 2, 3, \tag{4.24}$$

where ε_i is a positive constant referring to the speed of adjustment. Actually, this game model agrees well with the model proposed by Al-Khedhairi *et al.* in [94].

Based on system (4.24), we propose a new generalized model by introducing the Caputo-like difference operator into such a system. Specifically, the main focus of this section is to study the dynamics of the three bounded rationality firms with relative profit maximization and long memory of output decision. The new game model with the Caputo fractional order difference operator can be described as

$$^C\Delta_a^\nu q_1(t) = \varepsilon_1 q_1(t-1+\nu)(\alpha - \gamma_1 - 2q_1(t-1+\nu)$$
$$- (\zeta_1 - \zeta_2)q_2(t-1+\nu) - (\zeta_1 - \zeta_3)q_3(t-1+\nu)),$$
$$^C\Delta_a^\nu q_2(t) = \varepsilon_2 q_2(t-1+\nu)(\alpha - \gamma_2 - 2q_2(t-1+\nu)$$
$$- (\zeta_2 - \zeta_1)q_1(t-1+\nu) - (\zeta_2 - \zeta_3)q_3(t-1+\nu)),$$
$$^C\Delta_a^\nu q_3(t) = \varepsilon_3 q_3(t-1+\nu)(\alpha - \gamma_3 - 2q_3(t-1+\nu)$$
$$- (\zeta_3 - \zeta_1)q_1(t-1+\nu) - (\zeta_3 - \zeta_2)q_1(t-1+\nu)),$$
$$\tag{4.25}$$

where $\nu \in (0, 1]$ and $\zeta_i = \beta + \delta_i$. To study the dynamics of the three bounded rationalities with long memory, we need to define the discrete-time version of the game model. For this purpose, we need to replace a by zero and q_i by x_i, for each i. According to Theorem 1.11, we can gain the equivalent discrete-time formulas, defined as follows:

$$
\begin{cases}
x_1(n) = x_1(0) + \dfrac{1}{\Gamma(\nu)} \displaystyle\sum_{j=1}^{n} \dfrac{\Gamma(n-j+\nu)}{\Gamma(n-j+1)} [\varepsilon_1 x_1(j-1)(\alpha - \gamma_1 \\
\qquad\quad - 2x_1(j-1) - \theta_{12}x_2(j-1) - \theta_{13}x_3(j-1))], \\
x_2(n) = x_2(0) + \dfrac{1}{\Gamma(\nu)} \displaystyle\sum_{j=1}^{n} \dfrac{\Gamma(n-j+\nu)}{\Gamma(n-j+1)} [\varepsilon_2 x_2(j-1)(\alpha - \gamma_2 \\
\qquad\quad - 2x_2(j-1) - \theta_{12}x_1(j-1) - \theta_{23}x_3(j-1))], \\
x_3(n) = x_3(0) + \dfrac{1}{\Gamma(\nu)} \displaystyle\sum_{j=1}^{n} \dfrac{\Gamma(n-j+\nu)}{\Gamma(n-j+1)} [\varepsilon_3 x_3(j-1)(\alpha - \gamma_3 \\
\qquad\quad - 2x_3(j-1) - \theta_{13}x_1(j-1) - \theta_{23}x_2(j-1))],
\end{cases}
\tag{4.26}
$$

where $\theta_{ij} = \zeta_i - \zeta_j,\ \forall i, j = 1, 2, 3$.

4.8.2. *Stability Analysis*

For calculating the equilibrium points of the fractional order game model (4.25), we assign its left-hand side to zero as:

$$
\begin{cases}
\varepsilon_1 x_1(\alpha - \gamma_1 - 2x_1 - \theta_{12}x_2 - \theta_{13}x_3) = 0, \\
\varepsilon_2 x_2(\alpha - \gamma_2 - 2x_2 - \theta_{12}x_1 - \theta_{23}x_3) = 0, \\
\varepsilon_3 x_3(\alpha - \gamma_3 - 2x_3 - \theta_{13}x_1 - \theta_{23}x_2) = 0.
\end{cases}
\tag{4.27}
$$

By performing some algebraic computations, we can obtain the following fixed points:

$$F_1 = (0, 0, 0),\ F_2 = (0, 0.5(\alpha - \gamma_2), 0),\ F_3 = (0.5(\alpha - \gamma_1), 0, 0),$$

$$F_4 = (0, 0, 0.5(\alpha - \gamma_3)),$$

$$F_5 = \left(\frac{2(\gamma_1 - \alpha) + \theta_{12}(\alpha - \gamma_2)}{\theta_{12}^2 + 4}, \frac{2(\alpha - \gamma_2) + \theta_{12}(\alpha - \gamma_1)}{\theta_{12}^2 + 4}, 0 \right),$$

$$F_6 = \left(0, \frac{2(\alpha - \gamma_2) + \theta_{12}(\gamma_3 - \alpha)}{\theta_{23}^2 + 4}, \frac{2(\alpha - \gamma_3) + \theta_{23}(\alpha - \gamma_2)}{\theta_{23}^2 + 4}\right),$$

$$F_7 = \left(\frac{2(\alpha - \gamma_1) + \theta_{13}(\gamma_3 - \alpha)}{\theta_{13}^2 + 4}, 0, \frac{2(\alpha - \gamma_3) + \theta_{13}(\alpha - \gamma_1)}{\theta_{13}^2 + 4}\right),$$

$$F_8 = (A_1, A_2, A_3), \tag{4.28}$$

in which

$$A_1 = \frac{(\alpha - \gamma_3)(\theta_{12}\theta_{23} - 2\theta_{13}) - (\alpha - \gamma_2)(\theta_{13}\theta_{23} + 2\theta_{12}) + (\alpha - \gamma_1)(\theta_{23}^2 + 4)}{2(\theta_{13}^2 + \theta_{23}^2 + \theta_{12}^2 + 4)},$$

$$A_2 = \frac{-(\alpha - \gamma_3)(\theta_{12}\theta_{13} + 2\theta_{23}) + (\alpha - \gamma_2)(\theta_{13}^2 + 4) + (\alpha - \gamma_1)(-\theta_{13}\theta_{23} + 2\theta_{12})}{2(\theta_{13}^2 + \theta_{23}^2 + \theta_{12}^2 + 4)},$$

$$A_3 = \frac{(\alpha - \gamma_3)(\theta_{23}^2 + 4) + (\alpha - \gamma_2)(-\theta_{12}\theta_{13} + 2\theta_{23}) + (\alpha - \gamma_1)(\theta_{12}\theta_{23} + 2\theta_{13})}{2(\theta_{13}^2 + \theta_{23}^2 + \theta_{12}^2 + 4)}.$$

The Jacobian matrix of the fractional order game model (4.25) at an arbitrary point (x_1, x_2, x_3) can be outlined as

$$M(x_1, x_2, x_3) = \begin{pmatrix} D_1 - 2\varepsilon_1 x_1 & -\varepsilon_1\theta_{12}x_1 & -\varepsilon_1\theta_{13}x_1 \\ \varepsilon_2\theta_{12}x_2 & D_2 - 2\varepsilon_2 x_2 & -\varepsilon_2\theta_{23}x_2 \\ \varepsilon_3\theta_{13}x_3 & \varepsilon_3\theta_{23}x_3 & D_3 - 2\varepsilon_3 x_3 \end{pmatrix}, \tag{4.29}$$

where

$$D_1 = \varepsilon_1(\alpha - \gamma_1 - 2x_1 - \theta_{12}x_2 - \theta_{13}x_3),$$

$$D_2 = \varepsilon_2(\alpha - \gamma_2 - 2x_2 + \theta_{12}x_1 - \theta_{23}x_3),$$

$$D_3 = \varepsilon_3(\alpha - \gamma_3 - 2x_3 + \theta_{13}x_1 + \theta_{23}x_2).$$

To investigate the stability of the fractional order game system (4.25) at the fixed points reported above, we shall use the next theorem.

Theorem 4.1. [22] *Let x_f be a fixed point of the fractional order difference system $^C\Delta_a^\nu F(t) = F(x(t + \nu - 1))$, where $x(t) = (x_1(t), x_2(t), \ldots, x_n(t))^T$, and $J(x^f) = \frac{\partial f(x)}{\partial x}\Big|_{x=x^f}$ is the Jacobian matrix at the fixed point x_f, which is asymptotically stable if all the* *eigenvalues $(\lambda_i, \; i = 1, \ldots, n)$ of J satisfy:*

$$\lambda_i \in \left\{ z \in \mathbb{C} : |z| < \left(2\cos\frac{|\arg z| - \pi}{2 - \nu} \right)^{\nu} \; and \; |\arg z| > \frac{\nu\pi}{2} \right\},$$

$$\forall i = 1, \ldots, n. \tag{4.30}$$

In view of the previous discussion, we are now able to investigate the stability of the fixed points reported above. From this perspective, we introduce the next theoretical results.

Proposition 15. *The fixed point $F_1 = (0,0,0)$ is asymptotically stable if the fractional order value ν and the game parameters of map (4.25) satisfy*

$$\nu > \log_2 |\varepsilon_i(\alpha - \gamma_i)|, \quad and \quad \alpha < \gamma_i, \quad i = 1, 2, 3. \tag{4.31}$$

Proof. The Jacobian matrix (4.29) at the fixed point $F_1 = (0,0,0)$ can be easily computed as:

$$M_{F_1} = \begin{pmatrix} \varepsilon_1(\alpha - \gamma_1) & 0 & 0 \\ 0 & \varepsilon_2(\alpha - \gamma_2) & 0 \\ 0 & 0 & \varepsilon_3(\alpha - \gamma_3) \end{pmatrix}. \tag{4.32}$$

The associated characteristic equation is defined by:

$$(\varepsilon_1(\alpha - \gamma_1) - \lambda) \times (\varepsilon_2(\alpha - \gamma_2) - \lambda) \times (\varepsilon_3(\alpha - \gamma_3) - \lambda) = 0. \tag{4.33}$$

Consequently, the eigenvalues are:

$$\lambda_1 = \varepsilon_1(\alpha - \gamma_1), \; \lambda_2 = \varepsilon_2(\alpha - \gamma_2), \; \lambda_3 = \varepsilon_3(\alpha - \gamma_3).$$

Based on Theorem 4.1, it is easy to show that the fixed point $F_1 = (0,0,0)$ is always asymptotically stable when $\nu > \log_2 |\varepsilon_i(\alpha - \gamma_i)|$ and $\alpha < \gamma_i, \; \forall i = 1, 2, 3.$ $\qquad \square$

Proposition 16. *The conditions for asymptotic stability of the fixed point $F_2 = (0, 0.5(\alpha - \gamma_2), 0)$ are reported below:*

- $\alpha > \gamma_2$ *and* $\nu > \log_2 |\varepsilon_2(\gamma_2 - \alpha)|$.
- $\gamma_1 > \alpha + 0.5\theta_{12}(\gamma_2 - \alpha)$ *and* $\nu > \log_2 |\varepsilon_1(\alpha - \gamma_1 + 0.5\theta_{12}(\gamma_2 - \alpha))|$.
- $\gamma_3 > \alpha + 0.5\theta_{23}(\alpha - \gamma_2)$ *and* $\nu > \log_2 |\varepsilon_3(\alpha - \gamma_3 + 0.5\theta_{23}(\alpha - \gamma_2))|$.

Proof. The Jacobian matrix (4.29) at the fixed point $F_2 = (0, 0.5(\alpha - \gamma_2), 0)$ can be obtained as:

$$M_{F_1} = \begin{pmatrix} \varepsilon_1(\alpha - \gamma_1 - 0.5\theta_{12}(\alpha - \gamma_2)) & 0 & 0 \\ 0.5\theta_{12}\varepsilon_2(\alpha - \gamma_2)) & \varepsilon_2(\gamma_2 - \alpha) & 0 \\ 0 & 0 & \varepsilon_3(\alpha - \gamma_3)(1 + 0.5\theta_{23}) \end{pmatrix}. \tag{4.34}$$

Consequently, the eigenvalues are:

$$\lambda_1 = \varepsilon_1(\alpha - \gamma_1) + 0.5\varepsilon_1\theta_{12}(\gamma_2 - \alpha), \lambda_2 = \varepsilon_2(\gamma_2 - \alpha),$$

$$\lambda_3 = \varepsilon_3(\alpha - \gamma_3) + 0.5\varepsilon_3\theta_{23}(\alpha - \gamma_2).$$

It should be noted that these eigenvalues ensure condition (4.30) as reported in Theorem 4.1. □

In a similar manner to the proof of the previous two propositions, we can also derive the next results.

Proposition 17. *The conditions of asymptotic stability of the fixed point* $F_3 = (0.5(\alpha - \gamma_2), 0, 0)$ *are reported below:*

- $\alpha > \gamma_1$ *and* $\nu > \log_2 |\varepsilon_1(\gamma_1 - \alpha)|$.
- $\gamma_2 > \alpha + 0.5\theta_{12}(\alpha - \gamma_1)$ *and* $\nu > \log_2 |\varepsilon_2(\alpha - \gamma_2 + 0.5\theta_{12}(\alpha - \gamma_1))|$.
- $\gamma_3 > \alpha + 0.5\theta_{13}(\alpha - \gamma_1)$ *and* $\nu > \log_2 |\varepsilon_3(\alpha - \gamma_3 + 0.5\theta_{13}(\alpha - \gamma_1))|$.

Proposition 18. *The conditions for asymptotic stability of the fixed point* $F_4 = (0, 0, 0.5(\alpha - \gamma_2))$ *are reported below:*

- $\alpha > \gamma_3$ *and* $\nu > \log_2 |\varepsilon_3(\gamma_3 - \alpha)|$.
- $\gamma_1 > \alpha + 0.5\theta_{13}(\gamma_3 - \alpha)$ *and* $\nu > \log_2 |\varepsilon_1(\alpha - \gamma_1 + 0.5\theta_{13}(\gamma_3 - \alpha))|$.
- $\gamma_2 > \alpha + 0.5\theta_{23}(\gamma_3 - \alpha)$ *and* $\nu > \log_2 |\varepsilon_2(\alpha - \gamma_2 + 0.5\theta_{23}(\gamma_3 - \alpha))|$.

Proposition 19. *The conditions for asymptotic stability of the fixed point* $F_5 = \left(\frac{2(\gamma_1 - \alpha) + \theta_{12}(\alpha - \gamma_2)}{\theta_{12}^2 + 4}, \frac{2(\alpha - \gamma_2) + \theta_{12}(\alpha - \gamma_1)}{\theta_{12}^2 + 4}, 0 \right)$ *are reported below:*

- $\left[\alpha - \gamma_3 - \theta_{12} \frac{2(\gamma_1 - \alpha) + \theta_{12}(\alpha - \gamma_2)}{\theta_{12}^2 + 4} - \theta_{13} \frac{2(\alpha - \gamma_2) + \theta_{12}(\alpha - \gamma_1)}{\theta_{12}^2 + 4} \right] < 0$ *and*
  $\nu > \log_2 \left| \varepsilon_3 \left(\alpha - \gamma_3 - \theta_{12} \frac{2(\gamma_1 - \alpha) + \theta_{12}(\alpha - \gamma_2)}{\theta_{12}^2 + 4} - \theta_{13} \frac{2(\alpha - \gamma_2) + \theta_{12}(\alpha - \gamma_1)}{\theta_{12}^2 + 4} \right) \right|$.

- $-\frac{G}{2} \geq \sqrt{H}$ *and* $\nu > \log_2 \frac{\sqrt{|G^2 - 4H|} - G}{2}$ *where*

$$G = \varepsilon_1 \left(\alpha - \gamma_1 - 4 \frac{2(\gamma_1 - \alpha) + \theta_{12}(\alpha - \gamma_2)}{\theta_{12}^2 + 4} \right.$$

$$\left. - \theta_{12} \frac{2(\alpha - \gamma_2) + \theta_{12}(\alpha - \gamma_1)}{\theta_{12}^2 + 4} \right)$$

$$+ \varepsilon_2 \left(\alpha - \gamma_2 - 4 \frac{2(\alpha - \gamma_2) + \theta_{12}(\alpha - \gamma_1)}{\theta_{12}^2 + 4} \right.$$

$$\left. + \theta_{12} \frac{2(\gamma_1 - \alpha) + \theta_{12}(\alpha - \gamma_2)}{\theta_{12}^2 + 4} \right),$$

$$H = \varepsilon_1 \left(\alpha - \gamma_1 - 4 \frac{2(\gamma_1 - \alpha) + \theta_{12}(\alpha - \gamma_2)}{\theta_{12}^2 + 4} \right.$$

$$\left. - \theta_{12} \frac{2(\alpha - \gamma_2) + \theta_{12}(\alpha - \gamma_1)}{\theta_{12}^2 + 4} \right)$$

$$\times \varepsilon_2 \left(\alpha - \gamma_2 - 4 \frac{2(\alpha - \gamma_2) + \theta_{12}(\alpha - \gamma_1)}{\theta_{12}^2 + 4} \right.$$

$$\left. + \theta_{12} \frac{2(\gamma_1 - \alpha) + \theta_{12}(\alpha - \gamma_2)}{\theta_{12}^2 + 4} \right)$$

$$- \varepsilon_1 \varepsilon_2 \theta_{12}^2 \frac{2(\gamma_1 - \alpha) + \theta_{12}(\alpha - \gamma_2)}{\theta_{12}^2 + 4} \times \frac{2(\alpha - \gamma_2) + \theta_{12}(\alpha - \gamma_1)}{\theta_{12}^2 + 4}.$$

Proposition 20. *The conditions for asymptotic stability of the fixed point* $F_6 = \left(0, \frac{2(\alpha - \gamma_2) + \theta_{12}(\gamma_3 - \alpha)}{\theta_{23}^2 + 4}, \frac{2(\alpha - \gamma_3) + \theta_{23}(\alpha - \gamma_2)}{\theta_{23}^2 + 4} \right)$ *are reported below:*

- $\left[\alpha - \gamma_1 - \theta_{12} \frac{2(\alpha - \gamma_2) + \theta_{23}(\gamma_3 - \alpha)}{\theta_{23}^2 + 4} - \theta_{13} \frac{2(\alpha - \gamma_3) + \theta_{23}(\alpha - \gamma_2)}{\theta_{23}^2 + 4} \right] < 0$ *and*
$\nu > \log_2 \left| \varepsilon_1 \left(\alpha - \gamma_1 - \theta_{12} \frac{2(\alpha - \gamma_2) + \theta_{23}(\gamma_3 - \alpha)}{\theta_{23}^2 + 4} - \theta_{13} \frac{2(\alpha - \gamma_3) + \theta_{23}(\alpha - \gamma_2)}{\theta_{23}^2 + 4} \right) \right|.$

- $-\frac{I}{2} \geq \sqrt{J}$ *and* $\nu > \log_2 \frac{\sqrt{|I^2 - 4J|} - I}{2}$ *where*

$$I = \varepsilon_2 \left(\alpha - \gamma_3 - 4 \frac{2(\alpha - \gamma_2) + \theta_{23}(\gamma_3 - \alpha)}{\theta_{23}^2 + 4} - \theta_{23} \frac{2(\alpha - \gamma_3) + \theta_{23}(\alpha - \gamma_2)}{\theta_{23}^2 + 4} \right)$$

$$+ \varepsilon_3 \left(\alpha - \gamma_3 - 4 \frac{2(\alpha - \gamma_3) + \theta_{23}(\alpha - \gamma_2)}{\theta_{23}^2 + 4} + \theta_{23} \frac{2(\alpha - \gamma_2) + \theta_{23}(\gamma_3 - \alpha)}{\theta_{23}^2 + 4} \right),$$

$$J = \varepsilon_2 \left(\alpha - \gamma_3 - 4\frac{2(\alpha - \gamma_2) + \theta_{23}(\gamma_3 - \alpha)}{\theta_{23}^2 + 4} - \theta_{23}\frac{2(\alpha - \gamma_3) + \theta_{23}(\alpha - \gamma_2)}{\theta_{23}^2 + 4} \right)$$

$$+ \varepsilon_3 \left(\alpha - \gamma_3 - 4\frac{2(\alpha - \gamma_3) + \theta_{23}(\alpha - \gamma_2)}{\theta_{23}^2 + 4} + \theta_{23}\frac{2(\alpha - \gamma_2) + \theta_{23}(\gamma_3 - \alpha)}{\theta_{23}^2 + 4} \right)$$

$$- \varepsilon_3\varepsilon_2\theta_{23}^2 \frac{2(\alpha - \gamma_2) + \theta_{23}(\gamma_3 - \alpha)}{\theta_{23}^2 + 4} \times \frac{2(\alpha - \gamma_3) + \theta_{23}(\alpha - \gamma_2)}{\theta_{23}^2 + 4}.$$

Proposition 21. *The conditions for asymptotic stability of the fixed point* $F_7 = \left(\frac{2(\alpha - \gamma_1) + \theta_{13}(\gamma_3 - \alpha)}{\theta_{13}^2 + 4}, 0, \frac{2(\alpha - \gamma_3) + \theta_{13}(\alpha - \gamma_1)}{\theta_{13}^2 + 4} \right)$ *are reported below:*

- $\left[\alpha - \gamma_2 + \theta_{12}\frac{2(\alpha - \gamma_1) + \theta_{13}(\gamma_3 - \alpha)}{\theta_{13}^2 + 4} - \theta_{23}\frac{2(\alpha - \gamma_3) + \theta_{13}(\alpha - \gamma_2)}{\theta_{23}^2 + 4} \right] < 0$ *and*
  $\nu > \log_2 \left| \varepsilon_1 \left(\alpha - \gamma_1 - \theta_{12}\frac{2(\alpha - \gamma_2) + \theta_{23}(\gamma_3 - \alpha)}{\theta_{23}^2 + 4} - \theta_{13}\frac{2(\alpha - \gamma_3) + \theta_{13}(\alpha - \gamma_1)}{\theta_{23}^2 + 4} \right) \right|.$

- $-\frac{L}{2} \geq \sqrt{O}$ *and* $\nu > \log_2 \frac{\sqrt{|L^2 - 4O|} - L}{2}$ *where*

$$L = \varepsilon_2 \left(\alpha - \gamma_2 - 4\frac{2(\alpha - \gamma_1) + \theta_{13}(\gamma_3 - \alpha)}{\theta_{13}^2 + 4} - \theta_{23}\frac{2(\alpha - \gamma_3) + \theta_{13}(\alpha - \gamma_1)}{\theta_{13}^2 + 4} \right)$$

$$+ \varepsilon_3 \left(\alpha - \gamma_3 - 4\frac{2(\alpha - \gamma_3) + \theta_{13}(\alpha - \gamma_1)}{\theta_{13}^2 + 4} + \theta_{13}\frac{2(\alpha - \gamma_1) + \theta_{13}(\gamma_3 - \alpha)}{\theta_{13}^2 + 4} \right),$$

$$O = \varepsilon_2 \left(\alpha - \gamma_2 - 4\frac{2(\alpha - \gamma_1) + \theta_{13}(\gamma_3 - \alpha)}{\theta_{13}^2 + 4} - \theta_{23}\frac{2(\alpha - \gamma_3) + \theta_{13}(\alpha - \gamma_1)}{\theta_{13}^2 + 4} \right)$$

$$+ \varepsilon_3 \left(\alpha - \gamma_3 - 4\frac{2(\alpha - \gamma_3) + \theta_{13}(\alpha - \gamma_1)}{\theta_{13}^2 + 4} + \theta_{13}\frac{2(\alpha - \gamma_1) + \theta_{13}(\gamma_3 - \alpha)}{\theta_{13}^2 + 4} \right)$$

$$- \varepsilon_2\varepsilon_3\theta_{23}^2 \frac{2(\alpha - \gamma_1) + \theta_{13}(\gamma_3 - \alpha)}{\theta_{13}^2 + 4} \times \frac{2(\alpha - \gamma_3) + \theta_{13}(\alpha - \gamma_1)}{\theta_{13}^2 + 4}.$$

4.8.3. *Bifurcation Analysis and Numerical Simulations*

In this section, certain numerical experiments are simulated for the purpose of indicating the different route to chaos generated by the fractional order triopoly game system (4.25). The phase portraits, bifurcation diagrams and largest Lyapunov exponents of such systems are investigated for different levels of parameters and fractional order values. As explained in [94], there is a stable closed invariant curve around the Nash fixed point $F_8 = (0.8882, 0.4283, 0.0831)$. For instance, when selecting the parameters of the system under

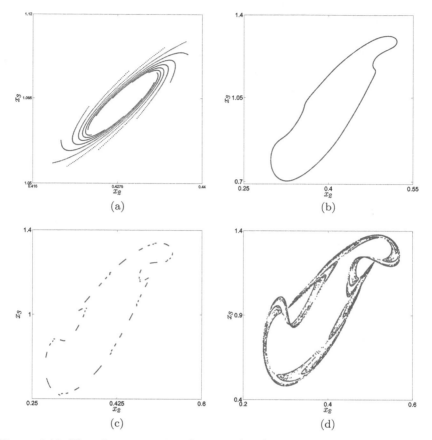

Figure 4.44. The phase portraits of system (4.25) with parameter values $\alpha = 2$, $\varepsilon_1 = 1.042811791$, $\varepsilon_2 = 1.1$, $\varepsilon_3 = 1.1$, $\zeta_1 = 0.4$, $\zeta_2 = 0.8$, $\zeta_3 = 0.1$, $\gamma_1 = 0.07$, $\gamma_2 = 0.03$, $\gamma_3 = 0.4$ for different fractional order values: (a) $\nu = 1$ (b) $\nu = 0.9$ (c) $\nu = 0.865$ (d) $\nu = 0.81$.

consideration as $\alpha = 2$, $\varepsilon_1 = 1.042811791$, $\varepsilon_2 = 1.1$, $\varepsilon_3 = 1.1$, $\zeta_1 = 0.4$, $\zeta_2 = 0.8$, $\zeta_3 = 0.1$, $\gamma_1 = 0.07$, $\gamma_2 = 0.03$, $\gamma_3 = 0.4$ and the initial values as $x_1(0) = 0.4$, $x_2(0) = 0.2$, $x_3(0) = 0.4$, then the stable closed invariant curve on $x_2 x_3$-plane is shown as in Figure 4.44(a). Besides, the closed invariant curve is expanded as in Figure 4.44. It can be seen that the closed invariant curve is affected by the fractional-order value ν, and a chaotic attractor is observed at $\nu = 0.81$. To further observe the dynamical behavior of system (4.25), its parameters are fixed as above and the fractional order value ν is

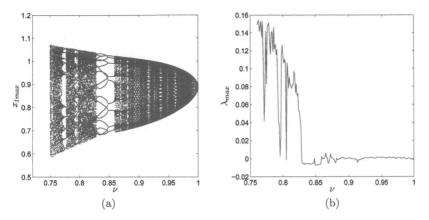

Figure 4.45. (a) The bifurcation diagram versus ν when $\alpha = 2$, $\varepsilon_1 = 1.042811791$, $\varepsilon_2 = 1.1$, $\varepsilon_3 = 1.1$, $\zeta_1 = 0.4$, $\zeta_2 = 0.8$, $\zeta_3 = 0.1$, $\gamma_1 = 0.07$, $\gamma_2 = 0.03$, $\gamma_3 = 0.4$. (b) The largest Lyapunov exponents with respect to ν corresponding to (a).

varied within the range $[0.72, 1]$. However, Figure 4.45 shows the bifurcation diagram and the largest Lyapunov exponents of the first player output $x_1(n)$. It can be shown that the long memory system begins from periodic states where the largest Lyapunov exponent equals zero, and then exhibits chaos at 0.8257. Figures 4.44 and 4.45 show the strong effect of the long memory on the stability of the equilibrium $F_8 = (0.8882, 0.4283, 0.0831)$. More precisely, the complexity of the model increases around the Nash fixed point and chaos appears as ν decreases. The strong effect of the long memory on the stability of the equilibrium $F_8 = (0.8882, 0.4283, 0.0831)$ can be seen in Figures 4.44 and 4.45. More precisely, the complexity of the model increases around the Nash fixed point, and chaos appears as ν decreases. However, Figure 4.46 shows the bifurcation diagram that is related to the adjustment parameter ε_1 when $\alpha = 2$, $\varepsilon_2 = 1.1$, $\varepsilon_3 = 1.1$, $\zeta_1 = 0.4$, $\zeta_2 = 0.8$, $\zeta_3 = 0.1$, $\gamma_1 = 0.07$, $\gamma_2 = 0.03$, $\gamma_3 = 0.4$. In particular, Figures 4.46(a) and 4.46(b) represent the bifurcation diagrams for $\nu = 0.985$ and $\nu = 0.972$, respectively. As one can see, the two diagrams are similar. Based on such figures, a stable Nash fixed point is observed as ε_1 increases from 1 to 1.02 when $\nu = 0.985$, whereas ε_1 increases from 1 to 1.007 when $\nu = 0.972$. We also observe that whenever we increase the value of ε_1, the Cournot

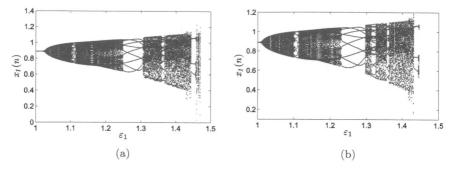

(a)   (b)

Figure 4.46. (a) The bifurcation diagram versus ε_1 with $\nu = 0.985$ when $\alpha = 2$, $\varepsilon_2 = 1.1$, $\varepsilon_3 = 1.1$, $\zeta_1 = 0.4$, $\zeta_2 = 0.8$, $\zeta_3 = 0.1$, $\gamma_1 = 0.07$, $\gamma_2 = 0.03$, $\gamma_3 = 0.4$. (b) The bifurcation diagram versus ε_1 with order $\nu = 0.972$.

Nash fixed point loses its stability via Neimark–Sacker bifurcation. Moreover, we observe that decreasing the fractional order value leads actually to the disappearance of the chaotic region. Besides, when $\nu = 0.985$, then system (4.25) exhibits chaotic behavior at $\varepsilon_1 \in [1.37, 1.443] \cup [1.461, 1.468]$. When $\nu = 0.972$, then the same system exhibits chaotic behavior for $\varepsilon_1 \in [1.38, 1.431]$. Therefore, it is possible to conclude initially that the speed of adjustment of the first player destablizes the dynamic of the market and make it unpredictable. For better observation, we discuss the chaos of fractional order triopoly game model (4.25) by fixing $\varepsilon_1 = 1.44$ and varying ν from 0.95 to 1 according to the step size $\Delta\nu = 0.6 \times 10^{-4}$. As a result, the bifurcation diagram of x_1 versus ν and the largest Lyapunov exponent diagram can be shown in Figure 4.47, illustrating that the states of the long memory system are different when ν decreases. In particular, when $\nu \in [0.9562, 0.9673] \cup [0.9809, 1]$, then model (4.25) is chaotic, where the maximum Lyapunov exponent is positive. At the same time, when $\nu \in [0.9562, 0.9809]$, the model is periodic. These results indicate that long memory can decrease the adjustment speed of the first firm, and the chaotic attractor is plotted in Figure 4.48 when $\nu = 0.98$. On the other hand, the periodic behavior of system (4.25) is illustrated in Figure 4.49 for $\nu = 0.975$, whereas the chaotic attractor, which is obtained when $\nu = 0.96$, is shown in Figure 4.50. Clearly, Figures 4.48–4.50 confirm the shape of

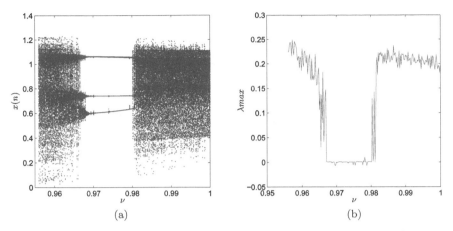

Figure 4.47. (a) The bifurcation diagram versus ν when $\alpha = 2$, $\varepsilon_1 = 1.44$, $\varepsilon_2 = 1.1$, $\varepsilon_3 = 1.1$, $\zeta_1 = 0.4$, $\zeta_2 = 0.8$, $\zeta_3 = 0.1$, $\gamma_1 = 0.07$, $\gamma_2 = 0.03$, $\gamma_3 = 0.4$. (b) The largest Lyapunov exponents with respect to ν corresponding to (a).

the bifurcation diagram exhibited in Figure 4.47(a) together with the plot of the largest Lyapunov exponent exhibited in Figure 4.47(b).

In [94], Al-kheidari *et al.* showed that, for the parameters $\alpha = 1$, $\varepsilon_2 = 0.9$, $\varepsilon_3 = 0.9$, $\zeta_1 = 0.4$, $\zeta_2 = 0.8$, $\zeta_3 = 0.1$, $\gamma_1 = 0.07$, $\gamma_2 = 0.03$, $\gamma_3 = 0.4$, there exists a flip bifurcation at $\varepsilon_1 = 2.305628076$ where the equilibrium $F_8 = (0.4451, 0.2384, 0.4502)$ loses its stability. The bifurcation diagram of the first player output $x_1(n)$ versus ε_1 is illustrated in Figure 4.51. Besides, Figure 4.51(a) shows that the Nash equilibrium point undergoes flip bifurcation and period-doubling route to chaos. By reducing the value of ν to 0.7635, we obtain the bifurcation diagram shown in Figure 4.51(b). It was found that the chaotic motion exists in the range $\varepsilon_1 \in [2.689, 3]$ with periodic windows at 2.873. The chaotic area increases when $\nu = 0.7635$ as shown in Figure 4.51(b). In order to confirm the shape of the bifurcation diagram in Figure 4.51(b), we investigate the system's behavior when $\varepsilon_1 = 2.6$ and $\varepsilon_1 = 2.9$. Namely, Figure 4.52 highlights the periodic behavior of system (4.25) obtained when $\varepsilon_1 = 2.6$. Based on these simulation results, one might deduce that the long memory increases the adjustment speed of the first player and the game loses its stability faster.

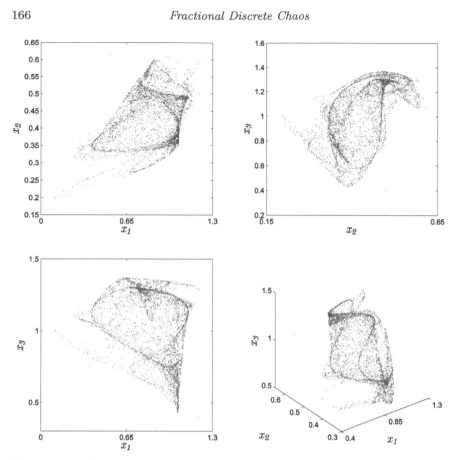

Figure 4.48. Chaotic attractor of system (4.25) with $\nu = 0.98$ and for $\alpha = 2$, $\varepsilon_1 = 1.44$, $\varepsilon_2 = 1.1$, $\varepsilon_3 = 1.1$, $\zeta_1 = 0.4$, $\zeta_2 = 0.8$, $\zeta_3 = 0.1$, $\gamma_1 = 0.07$, $\gamma_2 = 0.03$, $\gamma_3 = 0.4$.

4.8.4. *The 0–1 Test Method*

Here, the 0–1 test method is applied directly to the series data $x_1(n)$, which has been obtained from the first player, in order to study the influence of the fractional order value ν on the dynamics of the market. The results according to $\nu = 0.865$ and $\nu = 0.7635$ are shown in Figures 4.53 and 4.54, respectively. In particular, Figure 4.53 depicts bounded trajectories for $\nu = 0.865$, indicating that the suggested game is stable where the output $K = 0.000827$. On the other hand, the unbounded trajectories in Figure 4.54 confirm

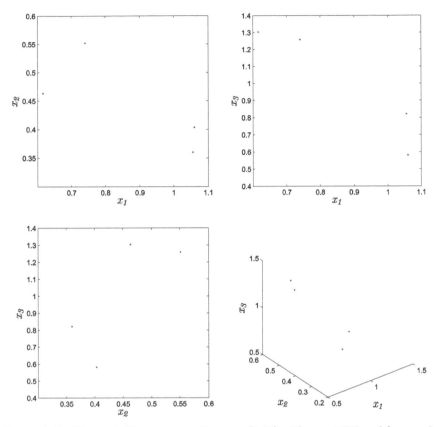

Figure 4.49. The periodic attractor of system (4.25) with $\nu = 0.975$ and for $\alpha = 2$, $\varepsilon_1 = 1.44$, $\varepsilon_2 = 1.1$, $\varepsilon_3 = 1.1$, $\zeta_1 = 0.4$, $\zeta_2 = 0.8$, $\zeta_3 = 0.1$, $\gamma_1 = 0.07$, $\gamma_2 = 0.03$, $\gamma_3 = 0.4$.

the chaotic behavior of the game for $\nu = 0.7635$ and the output $K = 0.995$. This clearly confirms the results reported in the previous section.

4.8.5. *Approximate Entropy*

In this section, we have applied the *ApEn* directly to the series of data $x_3(n)$ that has been obtained from the third firm. Figure 4.55 shows the approximate entropy of the fractional order Cournot game model (4.25) when $\alpha = 2$, $\varepsilon_1 = 1.042811791$, $\varepsilon_2 = 1.1$, $\varepsilon_3 = 1.1$, $\zeta_1 = 0.4$, $\zeta_2 = 0.8$, $\zeta_3 = 0.1$, $\gamma_1 = 0.07$, $\gamma_2 = 0.03$, and

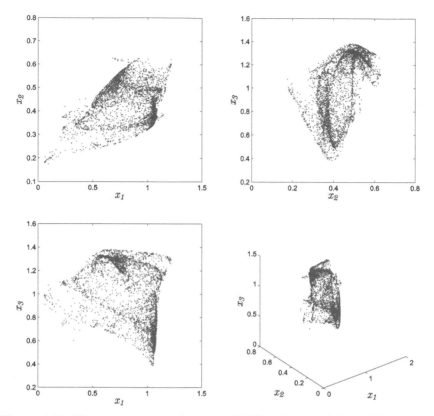

Figure 4.50. Chaotic attractor of system (4.25) with $\nu = 0.96$ and for $\alpha = 2$, $\varepsilon_1 = 1.44$, $\varepsilon_2 = 1.1$, $\varepsilon_3 = 1.1$, $\zeta_1 = 0.4$, $\zeta_2 = 0.8$, $\zeta_3 = 0.1$, $\gamma_1 = 0.07$, $\gamma_2 = 0.03$, $\gamma_3 = 0.4$.

$\gamma_3 = 0.4$, according to different fractional order values. Figure 4.55 shows that the results of the approximate entropy agree well with the corresponding bifurcation diagram and largest Lyapunov exponent as in Figure 4.45. It also shows that the smaller the fractional order value ν is, the more complex is the game model. Therefore, we must select the fractional order value in the game model (4.25) to have a relatively high structural complexity.

It is worth mentioning, at the end of this chapter, that the coexistence peculiarity that was previously discussed can be used for generating multipseudo signals. So the fractional order chaotic systems have potential applications in electronic measurement, secure

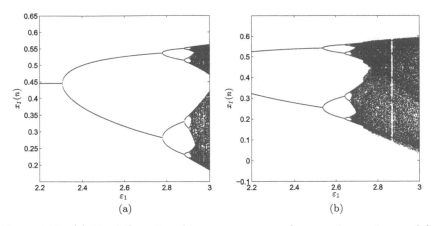

Figure 4.51. (a) The bifurcation diagram versus ε_1 when $\nu = 1$, $\alpha = 1$, $\varepsilon_2 = 0.9$, $\varepsilon_3 = 0.9$, $\zeta_1 = 0.4$, $\zeta_2 = 0.8$, $\zeta_3 = 0.1$, $\gamma_1 = 0.07$, $\gamma_2 = 0.03$, $\gamma_3 = 0.4$. (b) The bifurcation diagram versus ε_1 with $\nu = 0.7635$.

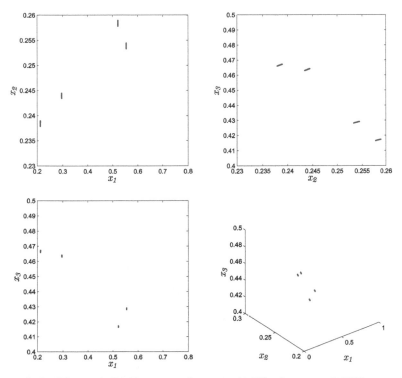

Figure 4.52. The periodic attractor of system (4.25) where $\nu = 0.7635$, $\varepsilon_1 = 2.6$, $\alpha = 1$, $\varepsilon_2 = 0.9$, $\varepsilon_3 = 0.9$, $\zeta_1 = 0.4$, $\zeta_2 = 0.8$, $\zeta_3 = 0.1$, $\gamma_1 = 0.07$, $\gamma_2 = 0.03$, $\gamma_3 = 0.4$.

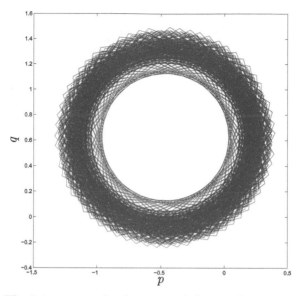

Figure 4.53. The 0–1 test: regular dynamics of the translation components (p, q) of the fractional order Cournot game model (4.25) when $\alpha = 2$, $\varepsilon_2 = 1.1$, $\varepsilon_3 = 1.1$, $\zeta_1 = 0.4$, $\zeta_2 = 0.8$, $\zeta_3 = 0.1$, $\gamma_1 = 0.07$, $\gamma_2 = 0.03$, $\gamma_3 = 0.4$, and $\nu = 0.865$.

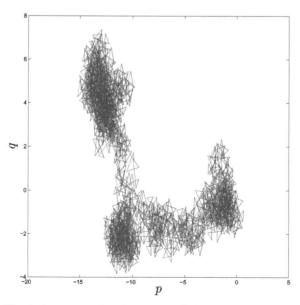

Figure 4.54. The 0–1 test: regular dynamics of the translation components (p, q) of the fractional order Cournot game model (4.25) when $\alpha = 2$, $\varepsilon_2 = 1.1$, $\varepsilon_3 = 1.1$, $\zeta_1 = 0.4$, $\zeta_2 = 0.8$, $\zeta_3 = 0.1$, $\gamma_1 = 0.07$, $\gamma_2 = 0.03$, $\gamma_3 = 0.4$, and $\nu = 0.7635$.

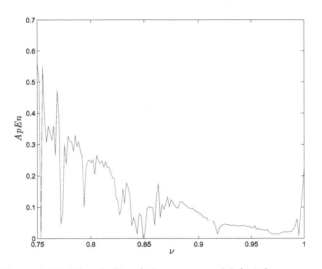

Figure 4.55. The *ApEn* of the game model (4.25) versus ν.

communications, information encryption and other fields. From this point of view, we confirm that the fractional order discrete-time extensions of chaotic maps can better describe the dynamics of these applications than that of integer order ones.

Chapter 5

Applications of Fractional Chaotic Maps

The analysis of fractional chaotic maps and their synchronization properties has attracted much attention over the past several years due to their potential applications in secure communication, encryption and other fields. In this chapter, we will present some applications of fractional chaotic maps. First, we will introduce different control laws to stabilize certain chaotic maps. Then, we will present some types of synchronization schemes for fractional discrete chaotic systems. Finally, we will illustrate an application of chaotic maps in implementing encryption and hardware.

5.1. Control of Fractional Chaotic Maps

In general, the capacity to control or stabilize chaotic dynamical systems is of particular interest. "Control" refers to adding adaptively updated terms to the chaotic system so that its states are close to zero. In this section, we introduce different linear and nonlinear controllers to stabilize certain fractional discrete chaotic systems.

Consider the following fractional map

$$^{C}\Delta_{\theta}^{\gamma} z(s) = g\left(z(s - 1 + \gamma)\right). \tag{5.1}$$

Let $C(s)$ be the adaptive controller. Then the controller fractional discrete system (5.1) is formulated as

$$^{C}\Delta_{\theta}^{\gamma} z(s) = g\left(z(s - 1 + \gamma)\right) + C(s - 1 + \gamma). \tag{5.2}$$

By adding control laws, the system becomes linear and can be written as

$$^{C}\Delta_{\theta}^{\gamma} z(s) = Bz(s - 1 + \gamma). \tag{5.3}$$

By using the stability conditions of a linear fractional discrete system given in Chapter 1 (Theorem 1.16), we can investigate that the zero fixed point of the system is asymptotically stable.

5.1.1. *Nonlinear Control Laws*

In the following, we propose two control laws related to a two-dimensional fractional map and a three-dimensional fractional discrete system. These control laws are taken from [51, 95].

5.1.1.1. *2D fractional map*

Consider the following fractional Duffing map [51]

$$\begin{cases} ^{C}\Delta_{\theta}^{\gamma} z_1(s) = z_2(s - 1 + \gamma) - z_1(s - 1 + \gamma), \\ ^{C}\Delta_{\theta}^{\gamma} z_2(s) = -\beta z_1(s - 1 + \gamma) + (\alpha - 1)z_2(s - 1 + \gamma) - z_2^3(s - 1 + \gamma). \end{cases} \tag{5.4}$$

For $\alpha = 2.77$ and $\beta = 0.2$, it has been shown that this map has chaotic behavior, which can be seen in the phase portraits presented in Figure 5.1.

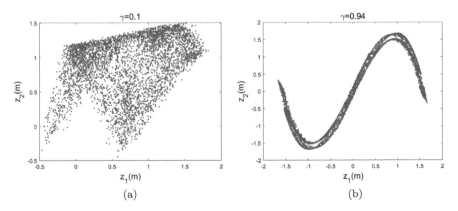

(a)   (b)

Figure 5.1. Phase portraits of the fractional Duffing map (5.4).

The control law for the fractional Duffing map (5.4) is presented in the following theorem.

Theorem 5.1. *The 2D fractional Duffing map (5.4) is stabilized subject to the following control law*

$$C(s) = \beta z_1(s) - \alpha z_2(s) + z_2^3. \tag{5.5}$$

Proof. The controlled fractional discrete system can be described as:

$$
\begin{cases}
{}^C\Delta_\theta^\gamma z_1(s) = z_2(s - 1 + \gamma) - z_1(s - 1 + \gamma), \\
{}^C\Delta_\theta^\gamma z_2(s) = -\beta z_1(s - 1 + \gamma) + (\alpha - 1)z_2(s - 1 + \gamma) \\
\qquad - z_2^3(s - 1 + \gamma) + C(s - 1 + \gamma).
\end{cases}
\tag{5.6}
$$

By substituting the control law (5.5) into (5.6), we get

$$
{}^C\Delta_\theta^\gamma (z_1(s - 1 + \gamma), z_2(s - 1 + \gamma))^T
$$
$$
= B(z_1(s - 1 + \gamma), z_2(s - 1 + \gamma))^T, \tag{5.7}
$$

where

$$
B = \begin{pmatrix} -1 & 1 \\ 0 & -1 \end{pmatrix}
$$

It is clear that the eigenvalues $\lambda_1 = -1$ and $\lambda_2 = -1$ of the matrix B satisfy the stability conditions of Theorem 1.19. So, the zero fixed point of the system (5.7) is asymptotically stable. $\quad\square$

In order to illustrate the result of Theorem 5.1, we take $\gamma = 0.94$, $(\alpha, \beta) = (2.77, 0.2)$ and the initial conditions $(z_1(0), z_2(0)) = (0.3, 0.1)$. The time evolution of the controlled fractional map is shown in Figure 5.2. Obviously, the states converge toward zero, which confirms the successful stabilization.

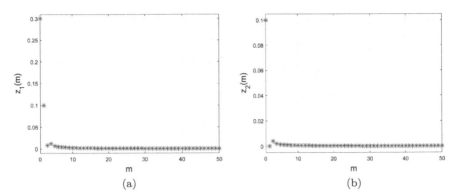

(a)   (b)

Figure 5.2. Time evolution of the states of the controlled fractional Duffing map.

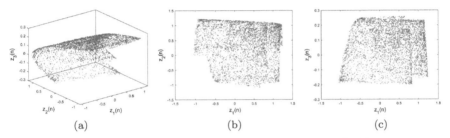

(a)   (b)   (c)

Figure 5.3. Phase portraits of the fractional Stefanski map (5.8) for $(\alpha, \beta) = (1.4, 0.2)$ and $\gamma = 0.96$.

5.1.1.2. *3D fractional map*

Consider the following fractional Stefanski map

$$
\begin{cases}
{}^{C}\Delta_{\theta}^{\gamma} z_1(s) = 1 + z_3(s - 1 + \gamma) - \alpha z_2^2(s - 1 + \gamma) - z_1(s - 1 + \gamma), \\
{}^{C}\Delta_{\theta}^{\gamma} z_2(s) = 1 + (\beta - 1)z_2(s - 1 + \gamma) - \alpha z_1^2(s - 1 + \gamma), \\
{}^{C}\Delta_{\theta}^{\gamma} z_2(s) = \beta z_1(s - 1 + \gamma) - z_3(s - 1 + \gamma),
\end{cases}
\tag{5.8}
$$

Reference [95] showed that there is chaotic behavior in this map for certain values of α and β, as can be seen in the phase portraits presented in Figure 5.3.

Theorem 5.2. *The 3D fractional discrete Stefanski system* (5.8) *is stabilized subject to the following control law*

$$\begin{cases} C_1(s) = \alpha z_2^2(s) - z_3 - 1. \\ C_2(s) = \alpha z_1^2(s) - 1. \end{cases} \tag{5.9}$$

Proof. The controlled fractional discrete system can be described as:

$$\begin{cases} {}^{C}\Delta_\theta^\gamma z_1(s) = 1 + z_3(s - 1 + \gamma) - \alpha z_2^2(s - 1 + \gamma) \\ \qquad\quad - z_1(s - 1 + \gamma) + C_1(s - 1 + \gamma), \\ {}^{C}\Delta_\theta^\gamma z_2(s) = 1 + (\beta - 1)z_2(s - 1 + \gamma) - \alpha z_1^2(s - 1 + \gamma) \\ \qquad\quad + C_1(s - 1 + \gamma), \\ {}^{C}\Delta_\theta^\gamma z_2(s) = \beta z_1(s - 1 + \gamma) - z_3(s - 1 + \gamma), \end{cases} \tag{5.10}$$

by substituting the control law (5.9) into (5.10), we get

$$ {}^{C}\Delta_\theta^\gamma z(s) = Bz(s - 1 + \gamma)^T, \tag{5.11}$$

where $z = (z_1, z_2, z_2)^T$ and

$$ B = \begin{pmatrix} -1 & 0 & 0 \\ 0 & \beta - 1 & 0 \\ \beta & 0 & -1 \end{pmatrix}. $$

It is not difficult to show that the eigenvalues of the matrix B satisfy the stability conditions of Theorem 1.19. So, the zero fixed point of the system (5.11) is asymptotically stable. $\square$

In order to verify this result, the states of the controlled fractional Stefanski map are shown in Figure 5.4. Clearly, the system is asymptotically stable and its states converge towards zero.

5.1.2. *Linear Control Laws*

Here, we present linear control laws to control chaotic fractional discrete systems defined by the Caputo h-difference operator. In addition, the Lyapunov direct method reported in Chapter 1

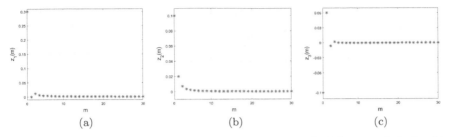

Figure 5.4. Time evolution of the states of the controlled fractional Stefanski map.

(Theorem 1.20) is used to prove the asymptotic convergence of the controllers. These control laws are taken from [96].

5.1.2.1. *2D fractional discrete system*

Consider the following fractional discrete system based on the Caputo h-difference operator [96]

$$\begin{cases} {}^{C}_{h}\Delta^{\gamma}_{\theta}z_1(s) = \beta z_1(s + h\gamma) - \beta z_1(s + h\gamma)z_2(s + h\gamma), \\ {}^{C}_{h}\Delta^{\gamma}_{\theta}z_2(s) = \beta \left(z_1^2(s - 1 + \gamma) - z_2(s + h\gamma) \right). \end{cases} \quad (5.12)$$

Taking the system parameter $\beta = 0.95$, the fractional order $\gamma = 0.9$ and the initial conditions $(z_1(0), z_2(0)) = (0.2, 0.3)$. Figure 5.5 illustrates the phase attractors of the fractional system (5.12). It can be observed that the states of the map are chaotic.

Theorem 5.3. *The 2D fractional discrete system* (5.12) *is stabilized subject to the following control law*

$$C(s) = -(1 + \beta)z_1(s), \quad s \in (h\mathbb{N})_{\theta + (1 - \gamma)h}. \quad (5.13)$$

Proof. The controlled fractional discrete system can be described as:

$$\begin{cases} {}^{C}_{h}\Delta^{\gamma}_{\theta}z_1(s) = \beta z_1(s + h\gamma) - \beta z_1(s + h\gamma)z_2(s + h\gamma) + C(s + h\gamma), \\ {}^{C}_{h}\Delta^{\gamma}_{\theta}z_2(s) = \beta \left(z_1^2(s - 1 + \gamma) - z_2(s + h\gamma) \right). \end{cases}$$

$$(5.14)$$

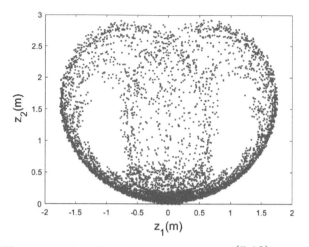

Figure 5.5. Phase portraits of the 2D fractional map (5.12) for $\beta = 0.95$ and $\gamma = 0.9$.

Substituting the control law (5.13) into (5.14), we obtain

$$\begin{cases} {}^{C}_{h}\Delta^{\gamma}_{\theta}z_1(s) = -z_1(s + h\gamma) - \beta z_1(s + h\gamma)z_2(s + h\gamma) \\ {}^{C}_{h}\Delta^{\gamma}_{\theta}z_2(s) = \beta \left(z_1^2(s - 1 + \gamma) - z_2(s + h\gamma) \right). \end{cases} \tag{5.15}$$

Now, consider the following Lyapunov function:

$$V(S) = \frac{1}{2}z_1^2(s) + \frac{1}{2}z_2^2(s), \tag{5.16}$$

This implies:

$${}^{C}_{h}\Delta^{\gamma}_{\theta}V(s) = \frac{1}{2}\,{}^{C}_{h}\Delta^{\gamma}_{\theta}z_1^2(s) + \frac{1}{2}\,{}^{C}_{h}\Delta^{\gamma}_{\theta}z_2^2(s).$$

Using Lemma 6, we obtain

$$\begin{aligned} {}^{C}_{h}\Delta^{\gamma}_{\theta}V &\leq z_1(s + h\gamma){}^{C}_{h}\Delta^{\gamma}_{\theta}z_1(s) + z_2(s + h\gamma){}^{C}_{h}\Delta^{\gamma}_{\theta}z_2(s) \\ &= -z_1^2(s + h\gamma) - \beta z_1^2(s + h\gamma)z_2(s + h\gamma) \\ &\quad + \beta z_1^2(s + h\gamma)z_2(s + h\gamma) - \beta z_2^2(s + h\gamma) \\ &= -z_1^2(s + h\gamma) - \beta z_2^2(s + h\gamma) < 0. \end{aligned}$$

Hence, according to Theorem 1.20, the system is asymptotically stable. $\qquad\qquad\square$

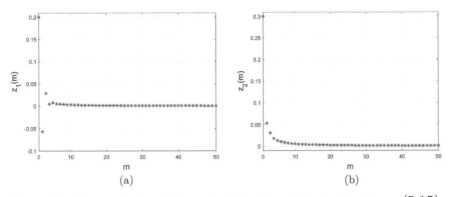

Figure 5.6. Time evolution of the controlled states of the fractional map (5.15).

To illustrate the theoretical results in the previous theorem, numerical simulations are performed. We choose $h = 1$, $\gamma = 0.9$ and the initial values $(z_1(0), z_2(0)) = (0.3, 0.1)$. Figure 5.6 reports the time evolution of the controlled states of h-fractional system (5.14). Clearly the errors converge to zero, which confirms the stabilization results.

5.1.2.2. *3D fractional discrete system*

Consider the following 3D fractional discrete system based on the Caputo h-difference operator [96]

$$
\begin{cases}
{}_h^C\Delta_\theta^\gamma z_1(s) = 1.4 + \beta z_3(s + h\gamma) + \alpha z_2(s + h\gamma) \\
\qquad - z_1^2(s + h\gamma) - z_1(s + h\gamma), \\
{}_h^C\Delta_\theta^\gamma z_2(s) = z_1(s - 1 + \gamma) - z_2(s + h\gamma). \\
{}_h^C\Delta_\theta^\gamma z_2(s) = z_2(s - 1 + \gamma) - z_3(s + h\gamma).
\end{cases}
\tag{5.17}
$$

Taking the system parameter $(\alpha, \beta) = (0.2, 0.1)$, the fractional order $\gamma = 0.98$ and the initial conditions $(z_1(0), z_2(0), z_3(0)) = (0.3, 0.3, 0.3)$. Figure 5.7 illustrates the phase attractors of the 3D fractional system (5.17). It can be observed that the states of the map are chaotic.

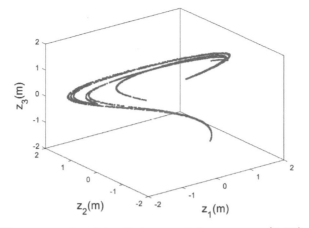

Figure 5.7. Phase portraits of the 3D fractional discrete map (5.12) for $\beta = 0.95$ and $\gamma = 0.9$.

Theorem 5.4. *The 3D fractional discrete system* (5.17) *is stabilized subject to the following control law*

$$\begin{cases} C_1(s) = -bz_1(s) - (\alpha + 1)z_2(s) - \beta z_3(s) - 1.4, \\ C_2(s) = -z_3(s), \end{cases} \quad (5.18)$$

where $|x(s)| \leq b$, $s \in (h\mathbb{N})_{\theta + (1-\gamma)h}$.

Proof. The controlled fractional discrete system can be described as:

$$\begin{cases} {}^{C}_{h}\Delta^{\gamma}_{\theta}z_1(s) = 1.4 + \beta z_3(s + h\gamma) + \alpha z_2(s + h\gamma) - z_1^2(s + h\gamma) \\ \qquad\qquad - z_1(s + h\gamma) + C_1(s + h\gamma), \\ {}^{C}_{h}\Delta^{\gamma}_{\theta}z_2(s) = z_1(s - 1 + \gamma) - z_2(s + h\gamma) + C_2(s + h\gamma). \\ {}^{C}_{h}\Delta^{\gamma}_{\theta}z_2(s) = z_2(s - 1 + \gamma) - z_3(s + h\gamma). \end{cases} \quad (5.19)$$

Substituting the control law (5.18) into (5.19), we obtain

$$\begin{cases} {}^{C}_{h}\Delta^{\gamma}_{\theta}z_1(s) = -z_2(s + h\gamma) - z_1^2(s + h\gamma) - (b + 1)z_1(s + h\gamma), \\ {}^{C}_{h}\Delta^{\gamma}_{\theta}z_2(s) = z_1(s - 1 + \gamma) - z_2(s + h\gamma) - z_3(s + h\gamma). \\ {}^{C}_{h}\Delta^{\gamma}_{\theta}z_2(s) = z_2(s - 1 + \gamma) - z_3(s + h\gamma). \end{cases} \quad (5.20)$$

Now, consider the following Lyapunov function:

$$V(S) = \frac{1}{2}z_1^2(s) + \frac{1}{2}z_2^2(s) + \frac{1}{2}z_3^2(s), \tag{5.21}$$

This implies:

$${}_{h}^{C}\Delta_{\theta}^{\gamma}V(s) = \frac{1}{2}\,{}_{h}^{C}\Delta_{\theta}^{\gamma}z_1^2(s) + \frac{1}{2}\,{}_{h}^{C}\Delta_{\theta}^{\gamma}z_2^2(s) + \frac{1}{2}\,{}_{h}^{C}\Delta_{\theta}^{\gamma}z_3^2(s).$$

Using Lemma 6, we obtain

$$
\begin{aligned}
{}_{h}^{C}\Delta_{\theta}^{\gamma}V &\leq z_1(s+h\gamma)\,{}_{h}^{C}\Delta_{\theta}^{\gamma}z_1(s) + z_2(s+h\gamma)\,{}_{h}^{C}\Delta_{\theta}^{\gamma}z_2^2(s) \\
&\quad + z_3(s+h\gamma)\,{}_{h}^{C}\Delta_{\theta}^{\gamma}z_3^2(s) \\
&= -z_1(s+h\gamma)z_2(s+h\gamma) - z_1^3(s+h\gamma) - (b+1)z_1^2(s+h\gamma) \\
&\quad + z_1(s+h\gamma)z_2(s+h\gamma) \\
&\quad - z_2^2(s+h\gamma) - z_2(s+h\gamma)z_3(s+h\gamma) \\
&\quad + z_2(s+h\gamma)z_3(s+h\gamma) - z_3^2(s+h\gamma) \\
&= -(b+1)z_1^2(s+h\gamma) - z_1^3(s+h\gamma) - z_2^2(s+h\gamma) - z_3^2(s+h\gamma) \\
&\leq -(b+1)z_1^2(s+h\gamma) + |z_1(s+h\gamma)|z_1^2(s+h\gamma) \\
&\quad - z_2^2(s+h\gamma) - z_3^2(s+h\gamma) \\
&\leq -(b+1)z_1^2(s+h\gamma) + bz_1^2(s+h\gamma) - z_2^2(s+h\gamma) - z_3^2(s+h\gamma) \\
&= -z_1^2(s+h\gamma) - z_2^2(s+h\gamma) - z_3^2(s+h\gamma) < 0.
\end{aligned}
$$

Hence, according to Theorem 1.20, the system is asymptotically stable. □

To illustrate the theoretical results in the previous theorem, numerical simulations are performed. Figure 5.8 reports the time evolution of the controlled states of h-fractional system (5.19). Clearly, the errors converge to zero, indicating that the above results are effective.

5.2. Synchronization in Fractional Chaotic Maps

Synchronization properties in fractional discrete systems represent an important research subject. The goal of synchronization is to force

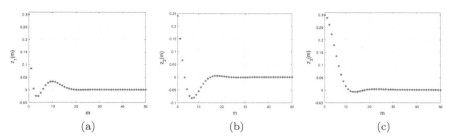

(a)   (b)   (c)

Figure 5.8. Time evolution of the controlled states of the fractional map (5.20), where (a) z_1 (b) z_2 and (c) z_3.

the synchronization error between the states of the master chaotic system and the slave chaotic systems to converge toward zero. In this section, by using some nonlinear controllers, we present different synchronization types of fractional order chaotic maps.

In the following, the master and slave systems are considered as follows

$$\begin{cases} {}^{C}\Delta_{\theta}^{\gamma}x(s) = F(x(s-1+\gamma)), & s \in \mathbb{N}_{\theta-\gamma+1}, \\ {}^{C}\Delta_{\theta}^{\gamma}z(s) = G(z(s-1+\gamma)) + C, \end{cases} \tag{5.22}$$

where $x(s) = (x_1(s), x_2(s), \ldots, x_n(s))^T \in \mathbb{R}^n$ is the master state vector and $z(s) = (Z_1(s), z_2(s), \ldots, z_m(s))^T \in \mathbb{R}^m$ is the slave state vector. $F : \mathbb{R}^n \to \mathbb{R}^n$, $G : \mathbb{R}^m \to \mathbb{R}^m$ are nonlinear functions and C is the vector controller to be determined.

5.2.1. *Generalized Synchronization (GS)*

Generalized synchronization is one of the most extensively researched types of synchronization. This type of synchronization is a generalization that includes complete synchronization, projective synchronization, anti-synchronization, and matrix projective synchronization.

Definition 17. If there exist a map $\Phi : \mathbb{R}^n \to \mathbb{R}^m$ and a controller C such that

$$\lim_{s \to \infty} \|e(s)\| = 0, \tag{5.23}$$

where $e(s) = z(s) - \Phi x(s)$ is the error system, then, the master–slave system (5.22) is synchronized under the generalized synchronization.

5.2.2. *Inverse Generalized Synchronization (IGS)*

Inverse generalized synchronization is the inversion of generalized synchronization. The error is calculated as the difference between the state of the master system and the function of the states of the slave system. This type of synchronization has been discussed in [97].

Definition 18. If there exist a map $\Psi : \mathbb{R}^m \to \mathbb{R}^n$ and a controller C such that

$$\lim_{s \to \infty} \|e(s)\| = 0, \tag{5.24}$$

where $e(s) = x(s) - \Psi z(s)$ is the error system, then, the master–slave system (5.22) is synchronized under the inverse generalized synchronization.

5.2.3. *Q–S Synchronization*

In the following, we discuss the Q–S synchronization of fractional discrete chaotic maps. The authors in [98] introduced this type of synchronization in 2005 for continuous-time dynamical systems, and then developed it to include discrete-time systems [99]. The Q–S synchronization of the fractional discrete-time system was studied in [100].

Consider the master and slave systems as follows

$$\begin{cases} {}^{C}\Delta_{\theta}^{\gamma} x(s) = Ax(s-1+\gamma) + F(x(s-1+\gamma)), & s \in \mathbb{N}_{\theta-\gamma+1}, \\ {}^{C}\Delta_{\theta}^{\mu} z(s) = Bz(s-1+\mu) + G(z(s-1+\mu)) + C, & s \in \mathbb{N}_{\theta-\mu+1}, \end{cases} \tag{5.25}$$

where $x(s) \in \mathbb{R}^n$ is the master state vector and $z(s) \in \mathbb{R}^m$ is the slave state vector. $A \in \mathcal{M}_n(\mathbb{R})$, $B \in \mathcal{M}_m(\mathbb{R})$. $F : \mathbb{R}^n \to \mathbb{R}^n$, $G : \mathbb{R}^m \to \mathbb{R}^m$ are nonlinear functions and C is the vector controller to be determined.

Definition 19. If there exist two functions $Q : \mathbb{R}^m \to \mathbb{R}^d$ and $Q : \mathbb{R}^m \to \mathbb{R}^d$ and a controller C such that

$$\lim_{s \to \infty} \|e(s)\| = 0, \tag{5.26}$$

where $e(s) = Q(z(s)) - S(x(s))$ is the error system, then, the master–slave system (5.25) is synchronized under the inverse Q–S synchronization.

5.2.4. *The Coexistence of Different Synchronization Types*

In the following, we present the coexistence of different types of synchronization for fractional discrete system [101]. Let us first define the types of synchronization of interest in this section.

Definition 20. If there exists a controller $C = (C_1, C_2, \ldots, C_m)$, then either constants $\nu \in \mathbb{R}^*$, a matrix $M = (m_{ij}) \in \mathbb{R}^{m \times n}$, a map $\Phi : \mathbb{R}^n \to \mathbb{R}^m$, a matrix $L = (L_{ij}) \in \mathbb{R}^{n \times m}$ or a map $\Psi : \mathbb{R}^m \to \mathbb{R}^n$ exist such that

$$\lim_{s \to \infty} \|z(s) - \nu x(s)\| = 0 \quad \Longrightarrow \quad \text{The pair (5.22) is PS.}$$
$$\lim_{s \to \infty} \|z(s) - \Phi x(s)\| = 0 \quad \Longrightarrow \quad \text{The pair (5.22) is GS.}$$
$$\lim_{s \to \infty} \|x(s) - \Psi z(s)\| = 0 \quad \Longrightarrow \quad \text{The pair (5.22) is IGS.}$$
$$\lim_{s \to \infty} \|z(s) - M x(s)\| = 0 \quad \Longrightarrow \quad \text{The pair (5.22) is FSHPS.}$$
$$\lim_{s \to \infty} \|x(s) - L z(s)\| = 0 \quad \Longrightarrow \quad \text{The pair (5.22) is IFSHPS.}$$

5.2.4.1. *Coexistence of PS, FSHPS and GS*

Consider the following 2D master system and 3D slave system

$$
\begin{cases}
{}^C\Delta_\theta^\gamma x_k(s) = F_k(x(s-1+\gamma)), \quad k = 1, 2, \\
{}^C\Delta_\theta^\gamma z_k(s) = \sum_{j=1}^{3} b_{kj} z_j(s-1+\gamma) + G_k(z(s-1+\gamma)) + C_k, \\
\qquad\qquad\qquad\qquad\qquad\qquad k = 1, 2, 3,
\end{cases}
$$

$$(5.27)$$

where $s \in \mathbb{N}_{\theta-\gamma+1}$, $\gamma \in (0, 1]$, $F_k : \mathbb{R}^2 \to \mathbb{R}$, $G_k : \mathbb{R}^3 \to \mathbb{R}$ are nonlinear functions, $B = (b_{kj}) \in \mathcal{M}_3(\mathbb{R})$ and C_k, $k = 1, 2, 3$, are the vectors controllers to be determined.

Definition 21. PS, FSHPS, and GS are said to coexist in the synchronization of the master–slave system (5.27) if there exists a controller $C = (c_1, c_2, c_3)^T$, a constant $\nu \in \mathbb{R}^*$, a constant matrix $M = (m_{kj}) \in \mathbb{R}^{1\times 2}$, and nonlinear map $\Phi : \mathbb{R}^2 \to \mathbb{R}$ such that the synchronization errors satisfy

$$\lim_{s\to\infty} \|e_k(s)\| = 0, \quad k = 1, 2, 3, \tag{5.28}$$

where

$$\begin{cases} e_1(s) = z_1(s) - \nu x_1(s), \\ e_2(s) = z_2(s) - M(x_1(s), x_2(s))^T, \\ e_3(s) = z_3(s) - \Phi(x_1(s), x_2(s)). \end{cases} \tag{5.29}$$

Theorem 5.5. *PS, FSHPS and GS coexist for the master–slave system* (5.27) *subject to*

$$\begin{cases} C_1(s) = \sum_{k=1}^{3}(d_{1k} - b_{1k})e_k(s) - \sum_{k=1}^{3}b_{1k}z_1(s) - G_1(z(s - 1 + \gamma)) \\ \qquad + \nu F_1(x(s - 1 + \gamma)), \\ C_2(s) = \sum_{k=1}^{3}(d_{2k} - b_{2k})e_k(s) - \sum_{k=1}^{3}b_{2k}z_2(s) - G_2(z(s - 1 + \gamma)) \\ \qquad + MF(x(s - 1 + \gamma)), \\ C_3(s) = \sum_{k=1}^{3}(d_{3k} - b_{3k})e_k(s) - \sum_{k=1}^{3}b_{3k}z_3(s) - G_3(z(s - 1 + \gamma)) \\ \qquad + {}^C\Delta_\theta^\gamma \Phi(x_1(s), x_2(s)), \end{cases} \tag{5.30}$$

where $D = (d_{ij}) \in \mathbb{R}^{3\times 3}$ *is a constant matrix chosen such that all the eigenvalues* λ_k *of* $B - A$ *satisfy* $-2^\gamma < \lambda_k < 0$, $k = 1, 2, 3$.

5.2.4.2. *Coexistence of IFSHPS and IGS*

Consider the following 2D master system and 3D slave system

$$
\begin{cases}
{}^{C}\Delta_\theta^\gamma x_k(s) = \sum_{j=1}^{2} a_{kj} x_j(s-1+\gamma) + F_k(x(s-1+\gamma)), & k=1,2, \\
{}^{C}\Delta_\theta^\gamma z_k(s) = G_k(z(s-1+\gamma)) + C_k, & k=1,2,3,
\end{cases}
$$

(5.31)

where $s \in \mathbb{N}_{\theta-\gamma+1}$, $\gamma \in (0,1]$, $F_k : \mathbb{R}^2 \to \mathbb{R}$, $G_k : \mathbb{R}^3 \to \mathbb{R}$ are nonlinear functions and $A = (a_{kj}) \in \mathcal{M}_2(\mathbb{R})$.

Definition 22. IFSHPS and IGS are said to coexist in the synchronization of the master–slave system (5.31) if there exists a controller $C = (c_1, c_2, c_3)^T$, a constant matrix $L = (l_{kj}) \in \mathbb{R}^{1 \times 3}$, and nonlinear map $\Psi : \mathbb{R}^3 \to \mathbb{R}$ such that the synchronization errors satisfy

$$
\lim_{s \to \infty} \|e_k(s)\| = 0, \quad k = 1,2, \tag{5.32}
$$

where

$$
\begin{cases}
e_1(s) = x_1(s) - L(z_1(s), z_2(s), z_3(s))^T, \\
e_2(s) = x_2(s) - \Psi(z_1(s), z_2(s), z_3(s)).
\end{cases}
\tag{5.33}
$$

Now, suppose that

$$
\Psi(z_1(s), z_2(s), z_3(s)) = \sum_{k=1}^{3} \eta_k z_k(s) + \phi(z_1(s), z_2(s), z_3(s)), \tag{5.34}
$$

where $\eta_1, \eta_2, \eta_3 \in \mathbb{R}$ and $\phi : \mathbb{R}^3 \to \mathbb{R}$ is a nonlinear function. Define

$$
R_1 = {}^{C}\Delta_\theta^\gamma x_1(s) - \sum_{k=1}^{3} l_k G_k(z(s-1+\gamma)), \tag{5.35}
$$

and

$$
R_2 = {}^{C}\Delta_\theta^\gamma x_2(s) - \sum_{k=1}^{3} l_k G_k(z(s-1+\gamma)) - {}^{C}\Delta_\theta^\gamma \phi(z_1(s), z_2(s), z_3(s)),
$$

(5.36)

then, the synchronization error can be written as

$$^C\Delta_\theta^\gamma e(s) = R - \Theta \times (C_1, C_2)^T - (l_3 C_3, \eta_3 C_3)^T, \qquad (5.37)$$

where $R = (R_1, R_2)^T$ and Θ is a matrix invertible such that

$$\Theta = \begin{pmatrix} l_1 & l_2 \\ \eta_1 & \eta_2 \end{pmatrix}$$

Theorem 5.6. *IFSHPS and IGS coexist for the master–slave system* (5.31) *subject to*

$$(C_1, C_2)^T = \Theta^{-1}[(K - A)e(s) + R] \quad and \quad C_3 = 0, \qquad (5.38)$$

where $K = (k_{ij}) \in \mathbb{R}^{2\times 2}$ *is a constant matrix chosen such that all the eigenvalues* λ_k *of* $A - K$ *satisfy* $-2^\gamma < \lambda_k < 0$, $k = 1, 2$.

5.3. Encryption Based on Fractional Discrete Chaotic Maps

It is well known that cryptography requires randomness and unpredictability, so it is clear that chaotic systems have good and suitable features in many applications related to encryption, owing to their sensitivity to initial conditions, periodicity, and reproducibility. In this section, we will present an example of the application of a fractional discrete chaotic map to the encryption of an electrophysiological signal. First, a pseudo-random bit generator (PRBG) will be designed based on the chaotic map values. Then, the electrophysiological signal will be combined with the fractional discrete chaotic map values in order to mask its structure. After that, to generate the encrypted signal, the modulated signal will be combined with a bitstream created by the PRBG. More details on this application are shown in [102].

5.3.1. *Design of Pseudo-Random Bit Generator (PRBG)*

Here, a PRBG is designed, having as its source of randomness the fractional discrete map. In order to take advantage of the fractional nature of the used map and cut down the computational cost, the map is implemented using finite memory, which is given as

$$z(m) = z(0) + \frac{1}{\Gamma(\gamma)} \sum_{j=1}^{L} \frac{\Gamma(m-j+\gamma)}{\Gamma(m-j+1)}$$

$$\times \left(A\sin\left(\frac{C}{z(m-L+j-1)}\right) - z(m-L+j-1) + B \right),$$

$$(5.39)$$

where L is the memory used. Here, until the full memory takes effect, we choose $L = 50$. Then in each iteration m of the fractional map (5.39), the bits are generated using the following rules:

$$b_{m,1} = \text{de2bi}\left(\text{mod} \left(\left\lfloor 10^{15} \left| \frac{\Gamma(m-1+\gamma)}{\Gamma(\gamma)\Gamma(m)} \right. \right. \right. \right.$$

$$\left. \left. \left. \left. \times \left(A\sin\left(\frac{C}{z(m-50)}\right) - z(m-50) + B \right) \right| \right\rfloor, 512 \right) \right),$$

$$b_{m,2} = \text{de2bi}\left(\text{mod} \left(\left\lfloor 10^{15} \left| \frac{\Gamma(m-2+\gamma)}{\Gamma(\gamma)\Gamma(m-1)} \right. \right. \right. \right.$$

$$\left. \left. \left. \left. \times \left(A\sin\left(\frac{C}{z(m-49)}\right) - z(m-49) + B \right) \right| \right\rfloor, 512 \right) \right),$$

$$\vdots$$

$$b_{m,50} = \text{de2bi}\left(\text{mod} \left(\left\lfloor 10^{15} \left| \frac{\Gamma(m-50+\gamma)}{\Gamma(\gamma)\Gamma(m-49)} \right. \right. \right. \right.$$

$$\left. \left. \left. \left. \times \left(A\sin\left(\frac{C}{z(m-1)}\right) - z(m-1) + B \right) \right| \right\rfloor, 512 \right) \right),$$

$$b_{m,51} = \text{de2bi}\left(\text{mod} \left(\left\lfloor 10^{15} |z(m)| \right\rfloor, 512 \right) \right), \qquad (5.40)$$

where $\lfloor . \rfloor$ represents the floor operation. So, each individual term of the sum in (5.39), for each state $z(m)$, is multiplied by 10^{15} in each iteration. The integer part of this product is taken modulo 512 and the result is transformed into its decimal representation. Thus, each of the terms $b_{m,1}, \ldots, b_{m,51}$ corresponds to 9 bits. The resulting stream of bits is the concatenation of the bits computed above:

$$\mathfrak{B} = \{b_{m-1,1}, \ldots, b_{m-1,51}, b_{m,1}, \ldots, b_{m,51}, \ldots\} \tag{5.41}$$

In general, there are $(L+1).9 = 459$ bits generated in each iteration. Thus, to reach a bitstream of length N, the fractional map needs to be iterated $\lfloor \frac{N}{459} \rfloor + 50$ times. So this approach utilizes the fractional structure of the map to counterbalance the computational burden in order to counterbalance the computational load for faster bit generation.

5.3.2. *Encryption of Electrophysiological Signal*

Now, we will present the encryption process. First, the structure of the source signal is modulated and obscured by combining it with the proposed chaotic map with parameters $z_1(0)$, A_1, B_1, and C_1. After that, the binary representation of the modulated signal is encrypted using the PRBG designed previously, constructed using a different chaotic map with different parameters $z_2(0)$, A_2, B_2, and C_2. The encryption process is carried out by using the bitwise XOR operator on the information and the chaotic bitstreams. Algorithm 1 describes the entire process.

In order to decrypt the signal, the receiver needs to know the key parameter values of the two chaotic maps used to generate the PRBG and modulate the signal. The key values of these maps are $\mathcal{K} = \{z_1(0), A_1, B_1, C_1, \gamma_1, z_2(0), A_2, B_2, C_2, \gamma_2\}$, so there are overall ten values. Assume that there is an accuracy of 16-digit available, then, the upper limit of the key space is $10^{10.16} = 10^{160} \approx (10^3)^{53.3} \approx (2^{10})^{53.3} = 2^{533}$. This is higher than the upper bound of 2^{100} required to resist brute force attacks [103]. Note that the actual key space is lower than 2533, since not all combinations of the key parameter values yield chaotic behavior.

Algorithm 1 Chaotic encryption of electrophysiological signal

---

    input: A signal ξ of length l.

        The key values of two chaotic maps of the form (5.39),

        $z_1(0)$, A_1, B_1, C_1, γ_1, $z_2(0)$, A_2, B_2, C_2, γ_2.

Output: An encrypted signal Eenc of the same length.

1. Modulate the signal to obscure its structure, as

$$\mathcal{M}_i = \xi_i + \frac{\lfloor 10^5 z_1(i) \rfloor}{10^3}, \quad i = 1, \ldots, l.$$

2. Transform the signal $\mathcal{M}$ into its binary representation $\mathcal{M}_{\text{bin}}$ of length $16 \times l$.

3. Generate a bitstream $\mathfrak{B}$ of length $16 \times l$ using the chaotic map $z_1(i)$

4. Encrypt the bitstream $\mathcal{M}_{\text{bin}}$ by combining it with the chaotic bistream as

$$\xi_{\text{bin}} = \mathcal{M}_{\text{bin}} \oplus \mathfrak{B}.$$

5. Transform the binary encrypted signal ξ_{bin} to decimal form, to obtain the encrypted signal ξ_{en}.

5.4. Electronic Implementation of Fractional Chaotic Maps

The hardware implementation of chaotic maps offers an efficient instrument for evaluating the theoretical model and concretely checking the existence of chaos (or hyperchaos) in the system dynamics. Electronic circuits such as switched-capacitor circuits [104] and microcontrollers [105, 106] can be used to realize a chaotic map. An example of a 2D fractional discrete chaotic map implemented using a microcontroller-based approach is presented in this section. A microcontroller-based approach was used due to its simplicity and low cost. This application is taken from [107].

5.4.1. *The Fractional Map with the Grunwald–Letnikov Operator*

Consider the following 2D discrete chaotic map with no fixed point:

$$\begin{cases} z_1(m+1) = z_2(m), \\ z_2(m+1) = \alpha_1 z_1(m) + \alpha_2 z_2(m) + \alpha_3 z_1^2(m) + \alpha_4 z_2^2(m) \\ \qquad + \alpha_5 z_1(m) z_2(m) + \alpha_6, \end{cases} \quad (5.42)$$

The fractional order version of this system (5.42) is described as follows

$$\begin{cases} {}^{GL}\Delta^{\gamma_1} z_1(m) = z_2(m) - z_1(m), \\ {}^{GL}\Delta^{\gamma_1} z_2(m) = \alpha_1 z_1(m) + \alpha_2 z_2(m) + \alpha_3 z_1^2(m) + \alpha_4 z_2^2(m) \\ \qquad + \alpha_5 z_1(m) z_2(m) + \alpha_6 - z_2(m), \end{cases} \quad (5.43)$$

where ${}^{GL}\Delta^{\gamma}$ denotes the Grunwald–Letnikov difference operator, which is defined as

$$ {}^{GL}\Delta^{\gamma} z(m) = \frac{1}{h^{\gamma}} \sum_{k=0}^{m} (-1)^k \binom{m}{k} z(m-k). \quad (5.44)$$

$\gamma > 0$ is the fractional order and $h \in]0, +\infty[$ is the sampling time.

Taking h as the unity, the GL difference operator (5.44) can be rewritten as:

$$ {}^{GL}\Delta^{\gamma} z(m) = z(m+1) - \gamma z(m) + \sum_{k=2}^{m+1} (-1)^k \binom{m}{k} z(m-k+1). $$

$$(5.45)$$

For simplicity, take $p = k - 1$, we obtain

$$ {}^{GL}\Delta^{\gamma} z(m) = z(m+1) - \gamma z(m) + \sum_{p=1}^{m} \beta_p z(m-p), \quad (5.46)$$

where $\beta_p = (-1)^{p+1} \binom{m}{p+1}$ is the binomial coefficient. By substituting Eq. (5.46) into the system (5.43), we obtain the numerical formula of the 2D fractional discrete map as follows

$$
\begin{cases}
z_1(m+1) = z_2(m) + (\gamma_1 - 1)z_1(m) - \displaystyle\sum_{p_1=1}^{m} \beta_{p_1} z(m-p_1), \\[2ex]
z_2(m+1) = \alpha_1 z_1(m) + \alpha_2 z_2(m) + \alpha_3 z_1^2(m) \\[1ex]
\qquad\qquad + \alpha_4 z_2^2(m) + \alpha_5 z_1(m)z_2(m) \\[2ex]
\qquad\qquad + \alpha_6 + (\gamma_2 - 1)z_2(m) - \displaystyle\sum_{p_2=1}^{m} \beta_{p_2} z(m-p_2),
\end{cases}
\tag{5.47}
$$

where β_{p_1} and β_{p_2} are the binomial coefficients, calculated using the following formula

$$
\beta_0 = -\gamma_i, \qquad \beta_{p_i} = \left(1 - \frac{\gamma_i + 1}{p_i + 1}\right)\beta_{p_i-1}, \quad i = 1, 2.
\tag{5.48}
$$

According to [108, 107], the binomial coefficient β_{p_i} decreases as the iteration p_i increases. Hence, a finite truncation length L is selected to approximate the iteration formula in order to facilitate the numerical computation. Therefore, for fixed L, the 2D fractional discrete model can be derived as

$$
\begin{cases}
z_1(m+1) = z_2(m) + (\gamma_1 - 1)z_1(m) - \displaystyle\sum_{p_1=1}^{L} \beta_{p_1} z(m-p_1), \\[2ex]
z_2(m+1) = \alpha_1 z_1(m) + \alpha_2 z_2(m) + \alpha_3 z_1^2(m) \\[1ex]
\qquad\qquad + \alpha_4 z_2^2(m) + \alpha_5 z_1(m)z_2(m) \\[2ex]
\qquad\qquad + \alpha_6 + (\gamma_2 - 1)z_2(m) - \displaystyle\sum_{p_2=1}^{L} \beta_{p_2} z(m-p_2),
\end{cases}
\tag{5.49}
$$

As one can see from Eq. (5.49), it is clear that the states of the 2D fractional discrete map are dependent on the previous L states, indicating that the concept of "short-memory" principle is introduced herein.

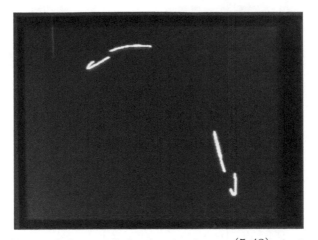

Figure 5.9. Chaotic behavior of the fractional map (5.43) obtained from the microcontroller-based approach.

5.4.2. *Hardware Implementation*

The 2D fractional discrete chaotic map (5.43) with no fixed point is realized by using the microcontroller-based approach. Setting the length $L = 20$, the initial conditions $z_1(0) = 0.51$, $z_2(0) = -5.04$, the parameters $\alpha_1 = 2.16$, $\alpha_2 = 0$, $\alpha_3 = 0.22$, $\alpha_4 = -0.22$, $\alpha_5 = 0.6$, $\alpha_6 = 0.76$ and the fractional orders $\gamma_1 = \gamma_2 = 0.99$. The microcontroller is presented with chaotic behavior in Figure 5.9. The experimental results clearly confirm the existence of chaos in the fractional discrete map, indicating the effectiveness of the proposed approach in the hardware implementation of the fractional discrete chaotic maps.

Bibliography

[1] Kelley, W. G., & Peterson, A. C. (2001). *Difference Equations: An Introduction with Applications* (Academic Press).

[2] Atici, F. M., & Eloe, P. W. (2007). A transform method in discrete fractional calculus. *Int. J. Diff. Eqs.* **2**(2).

[3] Atici, F., & Eloe, P. (2009). Initial value problems in discrete fractional calculus. *Proc. American Mathematical Society* **137**(3), 981–989.

[4] Miller, K. S., & Ross, B. Fractional difference calculus. *Proc. Int. Symp. Univalent Functions, Fractional Calculus and Their Applications*, Nihon University, Koriyama, Japan, May 1988, pp. 139–152 (Ellis Horwood).

[5] Abdeljawad, T. (2011). On Riemann and Caputo fractional differences. *Comput. Math. Appl.* **62**(3), 1602–1611.

[6] Holm, M. T. (2011). The theory of discrete fractional calculus: Development and application.

[7] Atici, F. M., & Sengul, S. (2010). Modeling with fractional difference equations. *J. Math. Anal. Appl.* **369**(1), 1–9.

[8] Abdeljawad, T., & Baleanu, D. (2011). Fractional differences and integration by parts. *J. Comput. Anal. Appl.* **13**(3).

[9] Abdeljawad, T. (2011). On Riemann and Caputo fractional differences. *Comput. Math. Appl.* **62**(3), 1602–1611.

[10] Abdeljawad, T. (2013). On delta and nabla Caputo fractional differences and dual identities. *Discrete Dynamics in Nature and Society.*

[11] Anastassiou, G. A. (2011). About discrete fractional calculus with inequalities. *Intelligent Mathematics: Computational Analysis* (Springer, Berlin, Heidelberg), pp. 575–585.

[12] Agarwal, R. P. (2000). *Difference Equations and Inequalities: Theory, Methods, and Applications* (CRC Press).

[13] Fulai, C., Xiannan, L., & Yong, Z. (2011). Existence results for nonlinear fractional difference equation. *J. Adv. Diff. Eqs.* **2011**(1), 1–12.

[14] Mozyrska, D., Girejko, E., & Wyrwas, M. (2013). Comparison of *h*-difference fractional operators. *Advances in the Theory and Applications of Non-integer Order Systems*, pp. 191–197.

[15] Mozyrska, D., & Girejko, E. (2013). Overview of fractional *h*-difference operators. *Advances in Harmonic Analysis and Operator Theory* (Birkhäuser, Basel), pp. 253–268.

[16] Ferreira, R. A., & Torres, D. F. (2011). Fractional *h*-difference equations arising from the calculus of variations. *Applicable Analysis and Discrete Mathematics*, 110–121.

[17] Elaydi, S. N. (2005). *An Introduction to Difference Equations* (Springer Nature, New York, NY).

[18] Mozyrska, D., & Wyrwas, M. (2015). The-transform method and delta type fractional difference operators. *Discrete Dynamics in Nature and Society.*

[19] Bohner, M., & Peterson, A. (2001). *Dynamic Equations on Time Scales: An Introduction with Applications* (Springer Science & Business Media).

[20] Holm, M. T. (2011). The Laplace transform in discrete fractional calculus. *Comput. Math. Appl.* **62**(3), 1591–1601.

[21] Kelley, W. G., & Peterson, A. C. (2001). *Difference Equations: An Introduction with Applications* (Academic Press).

[22] Cermák, J., Gyori, I., & Nechvátal, L. (2015). On explicit stability conditions for a linear fractional difference system. *Fract. Calculus and Appl. Anal.* **18**(3), 651.

[23] Abu-Saris, R., & Al-Mdallal, Q. (2013). On the asymptotic stability of linear system of fractional-order difference equations. *Fract. Calculus and Appl. Anal.* **16**(3), 613–629.

[24] Baleanu, D., Wu, G. C., Bai, Y. R., & Chen, F. L. (2017). Stability analysis of Caputo–like discrete fractional systems. *Commun. Nonlin. Sci. Numer. Simulat.* **48**, 520–530.

[25] Petras, I. (2011). *Fractional-order Nonlinear Systems: Modeling, Analysis and Simulation* (Springer Science & Business Media).

[26] Vulpiani, A., Cecconi, F., & Cencini, M. (2009). *Chaos: From Simple Models to Complex Systems*, Vol. 17 (World Scientific).

[27] Effah-Poku, S., Oben-Denteh, W., & Dontwi, I. K. (2018). A study of chaos in dynamical systems. *J. Math.*

[28] Eckmann, J. P., & Ruelle, D. (1985). Ergodic theory of chaos and strange attractors. *The Theory of Chaotic Attractors*, pp. 273–312.

[29] Ruelle, D. (Ed.). (1995). *Turbulence, Strange Attractors and Chaos*, Vol. 16 (World Scientific).

[30] Devaney, R. L., Siegel, P. B., Mallinckrodt, A. J., & McKay, S. (1993). A first course in chaotic dynamical systems: theory and experiment. *Comput. Phys.* **7**(4), 416–417.

[31] Banks, J., Brooks, J., Cairns, G., Davis, G., & Stacey, P. (1992). On Devaney's definition of chaos. *The American Mathematical Monthly* **99**(4), 332–334.

[32] Verhulst, P. F. (1845). Recherches mathématiques sur la loi d'accroissement de la population. *Journal des Économistes* **12**, 276.

[33] Hénon, M. A. (1976). Two-dimensional mapping with a strange attractor. *Comms. Math. Phys.* **50**, 69–77.

[34] Wu, G. C., & Baleanu, D. (2014). Discrete fractional logistic map and its chaos. *Nonlin. Dyn.* **75**(1), 283–287.

[35] Edelman, M., Macau, E. E., & Sanjuan, M. A. (eds.). (2018). *Chaotic, Fractional, and Complex Dynamics: New Insights and Perspectives* (Springer International Publishing).

[36] Peng, Y., Sun, K., He, S., & Wang, L. (2019). Comments on "Discrete fractional logistic map and its chaos" [*Nonlinear Dyn.* **75**, 283–287 (2014)]. *Nonlinear Dyn.* **97**(1), 897–901.

[37] Lyapunov, A. M. (1992). The general problem of the stability of motion. *Int. J. Contr.* **55**(3), 531–534.

[38] Oseledec, V. I. (1968). Multiplicative ergodic theorem.

[39] Benettin, G., Galgani, L., Giorgilli, A. *et al.* (1980). Lyapunov characteristic exponents for smooth dynamical systems and for hamiltonian systems; a method for computing all of them. Part 1: Theory. *Meccanica* **15**, 9–20.

[40] Benettin, G., Galgani, L., Giorgilli, A. *et al.* (1980). Lyapunov characteristic exponents for smooth dynamical systems and for hamiltonian systems; A method for computing all of them. Part 2: Numerical application. *Meccanica* **15**, 21–30.

[41] De Souza, S. L., & Caldas, I. L. (2004). Calculation of Lyapunov exponents in systems with impacts. *Chaos Solit. Fract.* **19**(3), 569–579.

[42] Eckmann, J. P., & Ruelle, D. (1992). Fundamental limitations for estimating dimensions and Lyapunov exponents in dynamical systems. *Physica D: Nonlin. Phenom.* **56**(2–3), 185–187.

[43] Wolf, A., Swift, J. B., Swinney, H. L., & Vastano, J. A. (1985). Determining Lyapunov exponents from a time series. *Physica D: Nonlin. Phenom.* **16**(3), 285–317.

[44] Vallejo, J. C., Sanjuan, M. A., & Sanjuan, M. A. (2017). *Predictability of Chaotic Dynamics* (Springer International Publishing).

[45] Geist, K., Parlitz, U., & Lauterborn, W. (1990). Comparison of different methods for computing Lyapunov exponents. *Progr. Theoret. Phys.* **83**(5), 875–893.

[46] Wu, G. C., & Baleanu, D. (2015). Jacobian matrix algorithm for Lyapunov exponents of the discrete fractional maps. *Commun. Nonlin. Sci. Numer. Simul.* **22**(1–3), 95–100.

[47] Wang, X. (2013). Calculation of negative Lyapunov exponents using a time series for potentially stable robotic systems.

[48] Parks, P. C., Thompson, J. M. T., & Stewart, H. B. (1986). *Non-Linear Dynamics and Chaos* (John Wiley & Sons, Chichester), 376.

[49] Bezziou, M., Dahmani, Z., Jebril, I., & Belhamiti, M. M. (2022). Solvability for a differential system of duffing type via Caputo–Hadamard approach. *Appl. Math. Inform. Sci.* **16**(2), 341–352.

[50] Dignowity, D., Wilson, M., Rangel-Fonseca, P., & Aboites, V. (2013). Duffing spatial dynamics induced in a double phase-conjugated resonator. *Laser Phys.* **23**(7), 075002.

[51] Ouannas, A., Khennaoui, A. A., Momani, S., & Pham, V. T. (2020). The discrete fractional duffing system: Chaos, 0–1 test, C_0 complexity, entropy, and control. *Chaos* **30**(8), 083131.

[52] Gottwald, G. A., & Melbourne, I. (2004). A new test for chaos in deterministic systems. *Proc. Royal Society of London. Series A: Mathematical, Physical and Engineering Sciences* **460**(2042), 603–611.

[53] Gottwald, G. A., & Melbourne, I. (2016). The 0–1 test for chaos: A review. *Chaos Detection and Predictability*, pp. 221–247.

[54] Gottwald, G. A., & Melbourne, I. (2009). On the implementation of the 0–1 test for chaos. *SIAM J. Appl. Dyn. Syst.* **8**(1), 129–145.

[55] Cafagna, D., & Grassi, G. (2008). Bifurcation and chaos in the fractional-order Chen system via a time-domain approach. *Int. J. Bifurcation and Chaos* **18**(07), 1845–1863.

[56] En-hua, S., Zhi-jie, C., & Fan-ji, G. (2005). Mathematical foundation of a new complexity measure. *Appl. Math. Mech.* **26**(9), 1188–1196.

[57] Cai, Z. J., & Sun, J. (2008). Modified C_0 complixity and applications. *J. Fudan Univ.* **47**(6), 791–799.

[58] He, S., Sun, K., & Wang, H. (2016). Solution and dynamics analysis of a fractional-order hyperchaotic system. *Math. Meth. Appl. Sci.* **39**(11), 2965–2973.

[59] Ran, J. (2018). Discrete chaos in a novel two-dimensional fractional chaotic map. *Adv. Diff. Eqs.* **2018**(1), 1–12.

[60] Ouannas, A., Khennaoui, A. A., Momani, S., Grassi, G., & Pham, V. T. (2020). Chaos and control of a three-dimensional fractional order discrete-time system with no equilibrium and its synchronization. *AIP Advances* **10**(4), 045310.

[61] Pincus, S. M. (1991). Approximate entropy as a measure of system complexity. *Proc. Nat. Acad. Sci.* **88**(6), 2297–2301.

[62] Pincus, S. (1995). Approximate entropy (ApEn) as a complexity measure. *Chaos* **5**(1), 110–117.

[63] Wang, C., & Ding, Q. (2018). A new two-dimensional map with hidden attractors. *Entropy* **20**(5), 322.

[64] Liu, Z., Xia, T., & Wang, J. (2018). Fractional two-dimensional discrete chaotic map and its applications to the information security with elliptic-curve public key cryptography. *J. Vibr. Contr.* **24**(20), 1–28.

[65] Jiang, H., Liu, Y., Wei, Z., & Zhang, L. (2016). A new class of three-dimensional maps with hidden chaotic dynamics. *Int. J. Bifurc. Chaos* **26**, 1650206.

[66] Panahi, S., Sprott, J. C., & Jafari, S. (2018). Two simplest quadratic chaotic maps without equilibrium. *Int. J. Bifurc. Chaos* **28**, 1850144.

[67] Hu, T. (2014). Discrete chaos in fractional Hénon map. *Appl. Math.* **5**(15), 2243–2248.

[68] Itoh, M., Yang, T., & Chua, L. (2001). Conditions for impulsive synchronization of chaotic and hyperchaotic systems. *Int. J. Bifurc. Chaos* **11**, 551–558.

[69] Elhadj, Z., & Sprott, J. C. (2008). A two-dimensional discrete mapping with C^∞ multifold chaotic attractors. *Electron. J. Theor. Phys.* **5**(17), 107–120.

[70] Sprott, J. C., & Xiong, A. (2015). Classifying and quantifying basins of attraction. *Chaos* **25**(8), 083101.

[71] Gasri, A., Ouannas, A., Khennaoui, A. A., Bendoukha, S., & Pham, V. T. (2020). On the dynamics and control of fractional chaotic maps with sine terms. *Int. J. Nonlin. Sci. Numer. Simul.* **21**(6), 589–601.

[72] Lu, J., Wu, X., Lu, J., & Kang, L. (2004). A new discrete chaotic system with rational fraction and its dynamical behaviors. *Chaos Solit. Fract.* **22**, 311–319.

[73] Rulkov, N. (2002). Modeling of spiking-bursting neural behavior using two-dimensional map. *Phys. Rev. E* **65**(4), 0419222.

[74] Chang, L., Lu, J., & Deng, X. (2005). A new two-dimensional discrete chaotic system with rational fraction and its tracking and synchronization. *Chaos Solit. Fract.* **24**, 1135–1143.

[75] Rulkov, N. F. (2002). Modeling of spiking-bursting neural behavior using two-dimensional map. *Phys. Rev. E* **65**(4), 041922.

[76] Ouannas, A., Khennaoui, A. A., Bendoukha, S., Wang, Z., & Pham, V. T. (2020). The dynamics and control of the fractional forms of some rational chaotic maps. *J. Syst. Sci. Compl.* **33**(3), 584–603.

[77] Chang, L., Lu, J., & Deng, X. (2005). A new two-dimensional discrete chaotic system with rational fraction and its tracking and synchronization. *Chaos Solit. Fract.* **24**, 1135–1143.

[78] Shivani, S., & Agarwal, S. (2016). VPVC: Verifiable progressive visual cryptography. *Patt. Anal. Appl.* **21**(1), 139–166.

[79] Goswami, A., Mukherjee, R., & Ghoshal, N. (2017). Chaotic visual cryptography based digitized document authentication. *Wireless Personal Commun.* **96**(3), 3585–3605.

[80] Merabet, N. E., & Benzid, R. (2018). Progressive image secret sharing scheme based on Boolean operations with perfect reconstruction capability. *Inform. Security J.: A Global Perspective* **27**(1), 14–28.

[81] Aziz-Alaoui, M. A., Robert, C., & Grebogi, C. (2001). Dynamics of a Hénon–Lozi-type map. *Chaos Solit. Fract.* **12**(12), 2323–2341.

[82] Ouannas, A., Khennaoui, A. A., Wang, X., Pham, V. T., Boulaaras, S., & Momani, S. (2020). Bifurcation and chaos in the fractional form of Hénon-Lozi type map. *The European Physical Journal Special Topics* **229**(12), 2261–2273.

[83] Zeraoulia, E., & Sprott, J. (2011). On the dynamics of a new simple 2D rational discrete mapping. *Int. J. Bifur. Chaos* **21**(1), 1–6.

[84] Zheng, J., Wang, Z., Li, Y., & Wang, J. (2018). Bifurcations and chaos in a three-dimensional generalized Hénon map. *Adv. Differ. Equ.* **2018**, 185.

[85] Gonchenko, S. V., Ovsyannikov, I. I., Simó, C., Turaev, D. V. (2005). Three-dimensional Hénon-like map and wild Lorenz-like attractors. *Int. J. Bifurc. Chaos* **15**(11), 3493–3508.

[86] Gonchenko, S. V., Gonchenko, V. S., & Tatjar, J. C. (2007). Bifurcation of three-dimensional diffeomorphisms non-simple quadratic homoclinic tangencies and generalized Hénon maps. *Regul. Chaotic Dyn.* **12**(3), 233–266.

[87] Gonchenko, S. V., Shilnikov, L. P., & Turaev, D. V. (2009). On global bifurcations in three-dimensional diffeomorphisms leading to wild Lorenz-like attractors. *Regul. Chaotic Dyn.* **14**(1), 137–147.

[88] Stefanski, K. (1998). Modelling chaos and hyperchaos with 3D maps. *Chaos Solit. Fract.* **9**, 83–93.

[89] Wang, X.Y. (2003). *Chaos in Complex Nonlinear Systems* (Publishing House of Electronics Industry, Beijing).

[90] Ouannas, A., Khennaoui, A. A., Grassi, G., & Bendoukha, S. (2019). On chaos in the fractional-order Grassi–Miller map and its control. *J. Comput. Appl. Math.* **358**, 293–305.

[91] Sun, H., Zhang, Y., Baleanu, D., Chen, W., & Chen, Y. (2018). A new collection of real world applications of fractional calculus in science and engineering. *Commun. Nonlin. Sci. Numer. Simul.* **64**, 213–231.

[92] Xin, B., Peng, W., & Kwon, Y. (2019). A fractional-order difference Cournot duopoly game with long memory. arXiv preprint arXiv:1903.04305.

[93] Cournot, A. (1963). Researches into the principles of the theory of wealth, Engl. Trans. *Irwin Paper Back Classics in Economics* (Chapter VII) (Hachette, Paris).

[94] Al-khedhairi, A. (2019). Differentiated Cournot duopoly game with fractional-order and its discretization. *Engineering Computations.*

[95] Khennaoui, A. A., Ouannas, A., Bendoukha, S., Grassi, G., Wang, X., Pham, V. T., & Alsaadi, F. E. (2019). Chaos, control, and synchronization in some fractional-order difference equations. *Adv. Diff. Eqs.* **2019**(1), 1–23.

[96] Talbi, I., Ouannas, A., Khennaoui, A. A., Berkane, A., Batiha, I. M., Grassi, G., & Pham, V. T. (2020). Different dimensional fractional-order discrete chaotic systems based on the Caputo h-difference discrete operator: dynamics, control, and synchronization. *Adv. Diff. Eqs.* **2020**(1), 1–15.

[97] Khennaoui, A. A., Ouannas, A., Bendoukha, S., Grassi, G., Wang, X., & Pham, V. T. (2018). Generalized and inverse generalized synchronization of fractional-order discrete-time chaotic systems with non-identical dimensions. *Adv. Diff. Eqs.* **2018**(1), 1–14.

[98] Yan, Z. (2005). QS (Lag or anticipated) synchronization backstepping scheme in a class of continuous-time hyperchaotic systems — a symbolic-numeric computation approach. *Chaos* **15**(2), 023902.

[99] Yan, Z. (2005). QS synchronization in 3D Hénon-like map and generalized Hénon map via a scalar controller. *Phys. Lett. A* **342**(4), 309–317.

[100] Ouannas, A., Khennaoui, A. A., Grassi, G., & Bendoukha, S. (2018). On the-chaos synchronization of fractional-order discrete-time systems: general method and examples. *Discrete Dynamics in Nature and Society* **2018**, 2950357.

[101] Bendoukha, S., Ouannas, A., Wang, X., Khennaoui, A. A., Pham, V. T., Grassi, G., & Huynh, V. V. (2018). The co-existence of different synchronization types in fractional-order discrete-time chaotic systems with non-identical dimensions and orders. *Entropy* **20**(9), 710.

[102] Gasri, A., Khennaoui, A. A., Ouannas, A., Grassi, G., Iatropoulos, A., Moysis, L., & Volos, C. (2022). A new fractional-order map with infinite number of equilibria and its encryption application. *Complexity* **2022**, 3592422.

[103] Alvarez, G., & Li, S. (2006). Some basic cryptographic requirements for chaos-based cryptosystems. *Int. J. Bifurc. Chaos* **16**(8), 2129–2151.

[104] Valtierra, J. L., Tlelo-Cuautle, E., & Rodríguez-Vázquez, Á. (2017). A switched-capacitor skew-tent map implementation for random number generation. *Int. J. Circuit The. Appl.* **45**(2), 305–315.

[105] Ablay, G. (2016). Chaotic map construction from common nonlinearities and microcontroller implementations. *Int. J. Bifurc. Chaos* **26**(7), 1650121.

[106] Huynh, V. V., Ouannas, A., Wang, X., Pham, V. T., Nguyen, X. Q., & Alsaadi, F. E. (2019). Chaotic map with no fixed points: entropy, implementation and control. *Entropy* **21**(3), 279.

[107] Ouannas, A., Khennaoui, A. A., Oussaeif, T. E., Pham, V. T., Grassi, G., & Dibi, Z. (2021). Hyperchaotic fractional Grassi–Miller map and its hardware implementation. *Integration* **80**, 13–19.

[108] Megherbi, O., Hamiche, H., Djennoune, S., & Bettayeb, M. (2017). A new contribution for the impulsive synchronization of fractional-order discrete-time chaotic systems. *Nonlin. Dyn.* **90**(3), 1519–1533.

Index